Collins discover

The Universe

Peter Grego

Collins

An Imprint of HarperCollins*Publishers*

ISBN-10: 0-00-721438-3
ISBN-13: 978-0-00-721438-9

ISBN-10: 0-06-089069-x (in the United States)
ISBN-13: 978-0-06-089069-8
FIRST U.S. EDITION. Published in 2006.

HarperCollins books may be purchased for educational, business, or sales promotional use. For information in the United States, please write to: Special Markets Department, HarperCollins Publishers, 10 East 53rd Street, New York, NY 10022.

The name of the "Smithsonian," "Smithsonian Institution," and the sunburst logo are registered trademarks of the Smithsonian Institution.

Created by **Focus Publishing**. Sevenoaks, Kent
Project editor: Guy Croton
Designer: David Etherington
Series design: Mark Thomson

Text © 2006, Peter Grego

Color reproduction by Colourscan, Singapore
Printed and bound by Printing Express Ltd, Hong Kong

10 09 08 07 06
6 5 4 3 2 1

Contents

1   **A big Universe**   6

2   **In cosmic realms**   22

3   **Third rock**   68

4   **Our cosmic backyard**   84

5   **The galactic neighborhood**   122

6   **Far and away**   152

7   **The Universe revealed**   172

Glossary   182

Need to know more?   186

Index   189

Acknowledgments   191

1 A big Universe

Ever since our ancestors were first struck by the majesty of the heavens and pondered its meaning, human beings have felt a profound curiosity about realms beyond the Earth. It is only when you start to investigate the position of even our closest planetary neighbors that you begin to realize the infinite other worlds that exist in the cosmos, at such immeasurable distances beyond our comprehension.

Splendor of the heavens

Although science has provided incredible insights into the Universe, our feelings of awe at viewing a star-spangled sky are undiminished by the knowledge that is now available to us.

Galaxies in their thousands—a deep view of the Universe, captured by the Hubble Space Telescope.

Inklings of the infinite

Throughout history, virtually everyone with the slightest degree of curiosity about the world around them has looked up at the sky and asked questions about it—seeking answers by questioning others, formulating their own theories, and by arriving at their own answers through observation.

Eternal musings

How far away are the stars and what's the farthest thing in the Universe that our eyes can see? Does it all go on forever? Are there other planets like the Earth, inhabited with beings intelligent enough to question who they are, and how they, their world, and the Universe itself came into being? If the Universe is infinite, could there be another person like me wondering exactly the same things at that very same moment in time? Will they reach the same conclusions as me? These sorts of questions have been asked by people of all ages ever since our ancestors developed the mental capacity to put aside their basic survival instincts for a moment or two and view the world around them—and the skies above them—with a genuine curiosity and a desire to know more about the cosmos.

Cosmic connection

The cosmos includes everything and while science has amassed a great deal of knowledge about the physical nature of our home planet, the intricate processes at work on its living surface, and in its oceans and atmosphere, remain only partly understood. In modern times, living in a polluted big city where the Sun itself competes for attention amid the concrete canyons, it is easy for an individual to feel utterly detached from the Earth and the rest of the Universe— spiritually, mentally, and physically. In the pre-industrial age, our ancestors were much more aware of the cycles of the heavens and the Earth. However, plant any 21st century urbanite beneath a dark, star-studded night sky, away from light pollution and the trappings of "civilisation," and all those hard-wired visceral feelings soon return.

Our physical bodies, our clothing and jewelry, the book you are holding right now and the chair you are sitting on— absolutely everything around you—is all made from material produced inside long-dead stars. Knowing that we are made of "star-stuff" allows us to feel more intimately connected with the Universe in which we exist.

Viewing the awesome cosmos.

Sunshine and starlight

Without the stars, the night skies would lose much of their splendor; without the Sun, there would be nobody around to appreciate the stars. The Sun may not be the most important star in the cosmos, but it is critical to the existence of the Earth.

Stellar energy, cosmic distances

That blindingly brilliant object that illuminates the daytime sky —our nearest star, the Sun—appears to be the single most important object in the heavens. If it emitted less heat and light than it currently does, humans would struggle to survive; our species would face the bleakest of prospects as the Earth's oceans froze and most plant and animal life on our planet became extinct. If its heat and light were switched off, human life could not survive at all. If the Sun were a solid lump of coal, it would burn itself to a cinder within a few thousand years. After millennia of speculation, the source of the Sun's prodigious output of energy (and that of all the stars visible in the sky) was finally explained in 1926 by the British scientist Arthur Eddington. Something far more powerful than simple chemical combustion powers the Sun. We rely on the thermonuclear processes elegantly encapsulated within Einstein's formula $E=mc^2$ for our continued existence.

As the skies gradually darken after sunset, stars begin to appear. It may seem ironic that the starlight upon which such hopes and dreams for the future are made actually set out on its journey across the galaxy in the remote past—its light may be just over four years old or more than 3,000 years old. Rigil Kent, a star in the constellation of Centaurus, is 4.4 light years away, and the bright star Deneb in Cygnus is 3,200 light years distant. Starlight takes 2.9 million light years to reach us from the nearest big galaxy to our own, the great spiral in Andromeda.

With the exception of the Sun, the stars are too far away to perceive as globes of glowing gas; even when viewed through a powerful telescope, they appear as mere pinpoints of light. Slight variation in color can be noticed between the stars—some appear white, others bluish, some slightly red. A star's color tells us a lot about its physical status—blue stars have intensely hot surfaces, while red stars have comparatively cool surfaces.

Light travel times to the earth can range from 1.3 light seconds (from the Moon) to 2.9 million light years (from the Andromeda Galaxy).

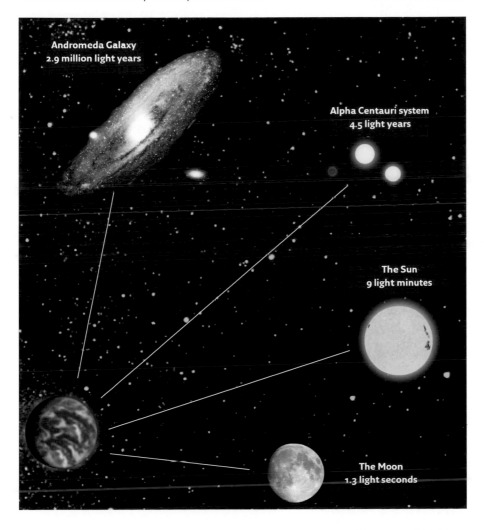

Andromeda Galaxy
2.9 million light years

Alpha Centauri system
4.5 light years

The Sun
9 light minutes

The Moon
1.3 light seconds

6.5 billion terranauts

We are all separated from the vacuum of space by just a few miles of breathable atmosphere. Earth spins on its axis once every 24 hours and revolves around the Sun once a year at a speed of more than 67,000 mph, accompanied by the Moon.

Big blue marble

From his vantage point in the command module of Apollo 8 above the Moon in December 1968, astronaut Jim Lovell described planet Earth as a "grand oasis in the big vastness of space." This grand oasis, a beautiful blue globe, two-thirds of whose surface is covered with water, measures 7,921 miles across. Our planet is the only place in the Universe known to have life. The fossil record shows that life has clung tenaciously to the Earth's surface for billions of years, surviving the

A crescent Earth rises over the rugged lunar surface, photographed by the crew of Apollo 17.

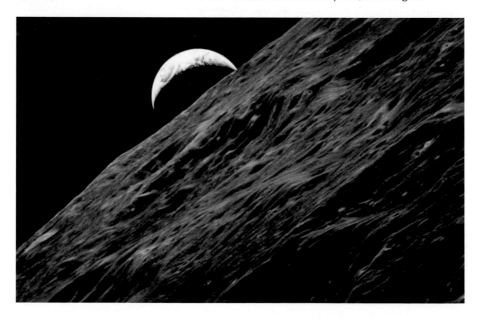

devastating effects of numerous major geological and cosmic catastrophes. Although mass extinctions of many species of animals have occurred, life—in one form or another—has carried on. Just a few dozen miles beneath the Earth's solid crust is a hot mantle—a zone of molten rock which has not cooled down since the Earth was born some 4.5 billion years ago. The Earth's crust occasionally splits in a volcanic eruption, allowing this material to erupt onto the surface.

Earth—the "grand oasis."

Moon musings

The same gravity that anchors you to the Earth keeps the Moon in its orbit. With a diameter about the same as the width of the continental United States, the Moon is in a near-circular orbit at an average distance of 238,713 miles from the Earth. It took the Apollo spacecraft a few days to traverse the gulf of space between the Earth and Moon, but it would take about a month to get there at the speed of a commercial jet airplane. The Moon is 400 times smaller than the Sun but it is 400 times closer to the Earth, so the Sun and Moon appear to be about the same size. Being a solid globe of rock, the Moon has no light of its own; moonshine is simply reflected sunlight. As the Moon orbits, the angle between the Moon, Earth, and Sun changes, causing a sequence of phases; the Moon appears to broaden from a narrow crescent in the evening skies to full Moon in around two weeks, and then becomes narrower again until it is a thin crescent in the morning skies. Sometimes, at new Moon, the Moon moves directly in front of the Sun, causing a solar eclipse; on the other side of its orbit, the full Moon occasionally enters the Earth's shadow. Eclipses only occur when the Earth, Moon, and Sun are in line.

Fellow wanderers

From little Mercury, whose rocky surface is scorched by the heat of the Sun, to countless deep-frozen cometary chunks from whose surface the Sun appears as a bright point, a remarkable collection of planets, asteroids, and comets makes up the Solar System.

Two planetary monarchs compared —Earth, queen of the terrestrial planets, and Jupiter, king of the gas giants.

Meet the neighbors

Our immediate planetary neighbors—Mercury, Venus, and Mars —are solid worlds like the Earth. Between Mars and the outer planets of the Solar System lies a zone occupied by countless chunks of rock. These remnants of the Solar System's formation range from the size of houses to the size of Iceland and are known as asteroids. Four giant planets preside over the outer reaches of the Solar System. Jupiter, Saturn, Uranus, and Neptune are all swathed in thick layers of mainly hydrogen gas. Jupiter, the largest gas giant, is so big that a thousand Earths could comfortably fit inside its vast volume. Pluto, the outermost planet, is a diminutive world, smaller in fact than our own Moon. More than four light hours from the Sun, Pluto is one of a number of icy worlds at the cold fringes of the observable Solar System. Far beyond the planets, clinging on to the Sun's gravity to a distance almost halfway to the nearest stars, lies an unseen realm of comets known as the Oort Cloud. Cometary visitors from this distant region occasionally speed through the inner Solar System; warmed by the Sun, their icy nuclei emit large amounts of gas and dust, producing celestial spectacles like the magnificent Comet Hyakutake of 1997.

The galactic suburbs

The Sun, attended by its nine major planets and their satellites, along with hundreds of thousands of asteroids and comets, orbits the center of a vast spiral galaxy of some 300 billion stars called the Milky Way. A middle-aged resident of the galactic suburbs, the Sun is located around 26,000 light years from the galactic center—about halfway from its center to the edge, within one of the galaxy's spiral arms. Orbiting the Milky Way at a speed of around 136 miles per second, the Sun has made around 20 galactic circuits since its birth around five billion years ago.

Measuring around four light years across (if we include the Oort Cloud), the Solar System occupies a tiny part of the Milky Way galaxy, some 100,000 light years in diameter. Indeed, the Sun would not be visible to the unaided eye from a distance of much more than 50 light years. Viewed from the Alpha Centauri system some 4.5 light years away, the Sun would appear as a bright star in the constellation of Cassiopeia.

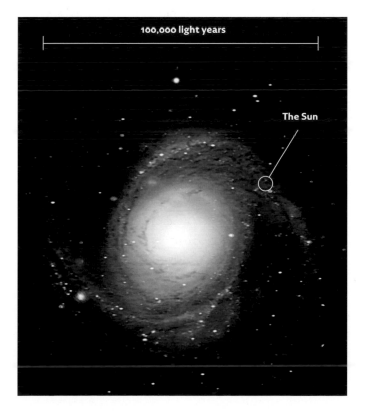

100,000 light years

The Sun

A hypothetical view from outside the Milky Way galaxy, showing the Sun's location.

Island Universe

On a clear night it's possible to see hundreds of stars from a rural location. Each one is a relatively nearby member of our home galaxy, whose farther reaches dissolve into a glowing band called the Milky Way.

Clouds of interstellar dust and gas in Sagittarius obscure our view of the Milky Way's bright core.

A grand design

Shaped like a flattened disk with a central bulge, our home galaxy the Milky Way is arranged in a loosely-wound spiral, with a number of curving arms composed of stars and glowing gas clouds. Two nearby small satellite galaxies—the Small and Large Magellanic Clouds—drift some distance beyond the edge of the Milky Way. From our perspective deep within an arm of the galaxy, we see the Milky Way as a glowing band which circles the heavens. Looking towards the bright galactic region in the vicinity of the constellation of Sagittarius, we are peering directly towards the center of the galaxy. Our view of the actual galactic hub is blocked by dense interstellar clouds of dust and gas, which show up as dark silhouettes against the brighter parts of the Milky Way.

Around 2.9 million light years away lies the Andromeda Galaxy—the nearest big galaxy to our own, the Milky Way.

Galaxies galore

As late as the early twentieth century most astronomers thought that the Milky Way represented the entire Universe. We now know that the Milky Way is just one of billions of other galaxies— some larger, some smaller than our own. The nearest big galaxy to the Milky Way is the Andromeda Galaxy, 2.9 million light years away, a spiral similar to our own. Discernible with the unaided eye on a dark, clear night, the light from that ghostly oval smudge in Andromeda set off long before the first sparks of consciousness flickered within the minds of human beings.

Telescopic surveys have shown us the structure of the wider Universe. Galaxies are arranged in gravitationally-bound clusters and superclusters, immersed in vast clouds of gas. Incredibly, the matter that can be observed telescopically—planets, stars, interstellar gas, and dust clouds—make up a small proportion of the matter in the Universe. A staggering 90 per cent is thought to take the form of "dark matter," currently unable to be detected by any telescope. This mysterious stuff is known to exist because its mass produces a detectable gravitational pull on galaxies. What constitutes dark matter is a subject of much debate among astronomers. It may be an entirely new form of matter, quite unlike the stuff we are made of, or it may simply be as yet unobserved ordinary matter such as old failed stars known as "brown dwarfs," or more exotic entities like black holes.

Zooming around the Universe

To get some sense of the size and scale of the cosmos, let's zoom away from the Earth in giant steps, right out to the very edge of the Universe, pausing to survey the scene before our eyes after each step.

Earth

Most of the Earth is contained within a 6,210 mile cube—just 1/30th light second across. From here we can see the blue oceans, the brilliant white clouds and icy polar caps, and the browns and greens of the continents. You may see some large sprawling gray cities during the day, but at night you won't find it difficult seeing city lights and illuminated roadways. From our high vantage point, the Earth's atmosphere—our protection against cosmic threats, such as bombardment by meteorites and harmful ultraviolet radiation from the Sun—appears awfully thin and insubstantial.

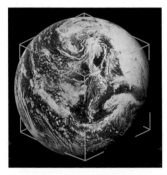

Our home planet, a sphere more than 7,880 miles across.

Earth-Moon

Virtually a double planet, the Earth-Moon system is contained within a cube 621,000 miles across. Light would take just 3 seconds to get from one side to the other. This view shows the comparative size and brightness of the Earth and the dull gray Moon. The Moon has no atmosphere; no clouds ever appear in the Moon's skies, rainfall never quenches the dry lunar soil, and no rivers flow on its surface. Our only natural satellite has always been dry and lifeless, its surface subject to harsh treatment from cosmic impactors.

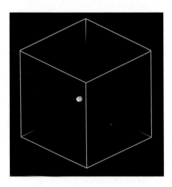

The little gray Moon is outshone by its big blue partner.

Solar System

Contained within a cube of space 6 billion miles on each side (8 light hours across) are the orbits of the nine major planets of the Solar System.

Oort Cloud

Viewing a cube of space 0.6 trillion miles (38 light days) on each side, all the major planets in the Solar System are lost in the glare of the Sun, which appears as bright as the full Moon. The inner parts of the distant Oort Cloud of comets are contained within the cube.

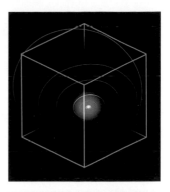

The planets are totally outshone by the brilliant Sun.

Stellar neighborhood

A cube measuring 100 x 100 x 100 light years centered on the Sun takes in a sizeable portion of the local stellar neighborhood, including many of the sky's brightest stars, from Rigil Kent (4.4 light years away) out to Capella (42 light years away).

Rigil Kent, otherwise known as Alpha Centauri, is the brightest star in the constellation of Centaurus. It is like the Sun in terms of age, size, color, and luminosity, but because it is so near to us it appears as the sky's fourth brightest star. Alpha Centauri is gravitationally bound to two other stars—a smaller orange star called Alpha Centauri B and a diminutive red dwarf known as Proxima Centauri. Of the three, Proxima is slightly the nearer; at 4.22 light years, it is the nearest star to the Sun. Could planets exist in orbit around Alpha Centauri? Since Alpha Centauri B orbits its brighter neighbor at a distance equivalent to the Sun and Uranus, Alpha Centauri wouldn't be able to hold on to any planets farther away from it as Jupiter is to the Sun, as their orbits would be gravitationally disrupted by Alpha Centauri B.

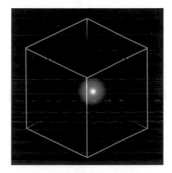

At the edge of the Sun's gravitational domain.

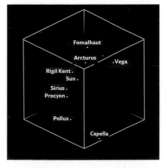

Fomalhaut
Arcturus
.Vega
Rigil Kent .
Sun .
Sirius .
Procyon .
Pollux .
Capella

Just an ordinary star among its stellar neighbors.

The spiral structure of our galaxy can only be appreciated when viewed from above its plane.

Milky Way

Our home galaxy, the vast spiral of the Milky Way, is enclosed within a cube 100,000 light years across. From this distance we can see only the brightest stars within the galaxy—stars of the Sun's brightness and dimmer are assumed within a mottled glow, interspersed with the dark silhouettes of opaque dust lanes and glowing clouds of dust and gas.

Our local group of galaxies.

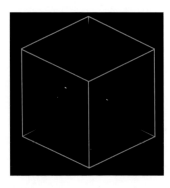

Local galactic group

A cube 10,000,000 light years across will take in the Local Group of galaxies. Only three large spirals dominate the scene—the Milky Way, the Andromeda Galaxy, and its neighbor the Triangulum Galaxy. The smaller galaxies dotted around are dwarf and irregular ones. Numerous gravitational interactions between the galaxies of our local group have taken place during the last 12 or more billion years.

Most of the Universe

From our impossible perspective, viewing a cube of space ten billion light years to each side, most of the objects in the visible Universe can be seen. Each dot represents a cluster of galaxies.

As we look farther out into space, we are looking backward in time. Light from the nearest galaxies in our local group takes a few hundred thousand to several million years to reach us, and we are seeing them as they appeared before *homo sapiens sapiens* walked the Earth. Galactic clusters and superclusters are arranged in filaments and sheets surrounding huge empty voids. Astronomers work out the distance of far away galaxies by measuring how much light is shifted towards the red end of the spectrum or redshifted. The greater the observed redshift, the greater a galaxy's distance. Telescopic deep field Images have revealed faint galaxies so distant that we see them as they appeared several billion years ago—before life itself developed on Earth.

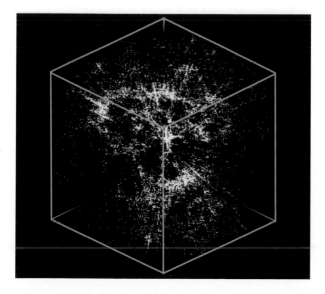

Our picture of the wider Universe is based on redshift measurements.

want to know more?

Take it to the next level...
▶ **Stellar distances** 63
▶ **Galactic redshifts** 64
▶ **Expanding Universe** 66
▶ **The Big Bang** 66

Other sources...
▶ **Computer software—** some programs give detailed data and images about near and deep space
▶ **Go observe! Invest in a pair of 7x50 binoculars and begin exploring the skies**

Weblinks...
▶ **Learn about the Solar System** www.nineplanets.org
▶ **Astronomy Picture of the Day—discover the cosmos** http://antwrp.gsfc. nasa.gov/apod/astropix .html
▶ **NASA's Imagine the Universe** http://imagine.gsfc. nasa.gov

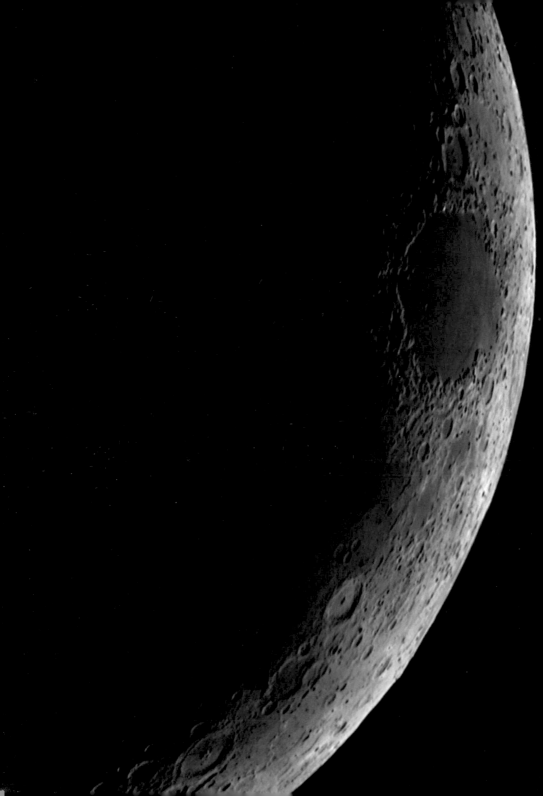

2 In cosmic realms

Humans have been watching the Universe for countless millennia. Records of celestial events go back several thousand years. Our attempts to understand the Universe and our place within it have occupied the thoughts of curious folk since the dawn of civilization. We take a look back at how our ancestors first interpreted what they saw in the skies above them and how that has evolved into the science we now use today.

Cosmic notions

Our distant ancestors certainly kept a watchful eye on celestial goings-on, for preserved in ancient cave paintings, rock carvings, and other human artefacts are indisputable records of a variety of astronomical events.

Observant ancestors

Patterns of dots thought to be star maps and what appears to be an attempt to populate the heavens with figures representing constellations in the form of animals have been discovered at a number of prehistoric sites in various parts of the world. They show an amazing sophistication of thought in our distant ancestors, proving that they were immensely observant of the world around them and the skies above their heads.

Wonderful scenes like this— brilliant Venus and ruddy Mars in Scorpius, rising over the French Alps—have been witnessed by human eyes for countless millennia.

Moon markings

One of the most familiar symbols to be found is the crescent representing the Moon and markings that indicate the monthly lunar cycle of phases. One such lunar calendar of 29 markings—one for each day of the lunar month—has recently been identified in the caves at Lascaux in France, dated to around 15,000 BC. This symbolism, acknowledging the Moon's importance, is hardly surprising; the Moon's light will have been of immense value to any culture without the benefit of streetlights and flashlights, in terms of hunting, nocturnal survival, and in primitive religions which attached significance to Moon worship and observing the lunar cycles. What many experts consider to be a depiction of the Moon showing the main features discernible with the unaided eye—the dark patches, known as the "maria"—has been discovered in 5,000-year-old rock carvings in prehistoric tombs at Knowth, County Meath in Ireland.

Celestial purpose

Tens of thousands of years ago our ancestors believed the skies had a greater purpose than to be simply admired for their splendor. The heavens were considered to have a purpose; those fixed stellar points of light actually meant something. Those bright objects which appeared to move—the Moon, Sun, and the five planets—were imagined to have a special significance. So the skies became incorporated into human affairs, and astrology was born. Unexpected changes in the skies—bright comets, eclipses, and novae—were considered to be potent warnings or revelations to humanity.

Carved into Irish rock 50 centuries ago, this set of curves found in the Knowth prehistoric tomb in Ireland may represent the oldest map of the Moon.

Astrology

The notion that the movements of the Sun, Moon, and five planets have an influence on the lives of individual people and the course of world events has its roots in ancient concepts and beliefs.

Predictive power

Being able to predict celestial events, from the movements of Venus to eclipses of the Sun and the Moon, appeared to give a great advantage to any society capable of mastering such complex astronomical and mathematical problems. Great civilizations, such as those of ancient Babylon, Egypt, and China, attached great importance to observing, recording, and predicting heavenly phenomena. Astrologer-priests kept a constant vigil on the skies, ostensibly for society's well-being and to keep their rulers informed of any celestial portents that might affect the status quo. Astrology was considered such a precious asset that its use without the ruler's permission was often punishable by execution. Sometimes, either through lack of attention to the skies, sloppy mathematics, or just plain bad luck, the astrologers got things wrong—and paid a severe penalty as a result. The ancient Chinese *Book of History* reports that two court astrologers were executed for having failed to announce a total lunar eclipse in 2136 BC.

Comets were once regarded as celestial omens.

Portents of doom

Of course, not everything in the heavens can be predicted. Even these days, bright novae or supernovae (stars which temporarily brighten far beyond their ordinary brightness), bright comets, fireballs (brilliant meteors), or aurorae (displays of the northern/southern lights) often take astronomers by complete surprise. Although such sights can be awesome to behold, we know how these phenomena are caused and realize

Astrology may look glamorous, but it is as insubstantial as the paper its predictions are written on.

that, for the most part, they pose no threat to the Earth or its inhabitants. Only a few hundred years ago, things were very different, and a bright comet might have been regarded as an omen of impending change on the Earth—of natural catastrophe, famine, pestilence, war, or a change of ruler.

Pseudoscience

Ancient astrology was practiced by making astronomical observations (in an age long before the telescope was invented), so the two subjects of astrology and astronomy were once closely intertwined. Modern forms of astrology—which claim to foretell the destinies of individuals—have often come under close scientific scrutiny, yet have been consistently found to have no provable predictive powers. While many find modern astrology to be entertaining, it has no basis in science, apart from using certain scientific terms and nomenclature.

Ancient cosmology

Although humans around the world have viewed the same range of celestial phenomena through the ages, our interpretation of them—how they were thought to have been created and what they were believed to have meant—has been immensely varied.

A copy of observations of Venus made in Babylon in 1700 BC is recorded on this clay tablet made at Nineveh in the 7th century BC.

Mesopotamian skies

The science of astronomy can trace its roots back thousands of years to Mesopotamia, a land bordered by the rivers Tigris and Euphrates in modern Iraq.

Fertile minds

Mesopotamia actually means "land between two rivers." The region is also known as "the fertile crescent," its fertility being perhaps the source of the biblical Garden of Eden. Here, from around 4000 BC, the first complex human civilizations grew: first Sumeria, then the kingdoms of Babylon and Assyria.

Omens in the heavens were deemed of tremendous importance to the rulers of ancient Mesopotamia—all events visible in the skies were believed to take place for a reason, and it was the job of the astronomer-priests to note such events, predict them where possible, and to interpret their meaning. Astronomical observations were deemed so important that they were preserved on clay tablets imprinted with cuneiform script—a form of writing made with narrow wooden scribes with tapered ends.

A flourishing sky lore also developed in Sumeria, in which ancient myths and legends were projected into the heavens, creating some of the constellations with which we are familiar today. Zodiacal constellations such as Sagittarius, Scorpius, Capricornus, Leo, Gemini, and Taurus—areas of the sky through which the Sun, Moon, and five planets were observed to travel—were devised by the Sumerians around 5,000 years ago. In addition to having a spiritual significance, the constellations had a practical use—their visibility throughout the year (notably the times of the year when certain constellations were first seen rising or setting), was used to mark agricultural seasons.

Lunar calendar

A calendar based on the Moon's cycles was used. Each month began with the first sighting of the narrow crescent Moon at sunset, and 12 lunar months made one year. Since 12 lunar months is 11 days short of a solar year (365 days), the Sumerian calendar was synchronized with the solar year by adding a leap month every three or four years.

Babylonian astronomy

The Babylonians adopted and added to ancient Sumerian sky lore, their calendar, and scientific knowledge. Hundreds of Babylonian clay tablets noting scientific and astronomical subjects have been discovered. The Babylonians acknowledged the "morning star" and the "evening star" to be a single object—the planet Venus. Lunar eclipses were capable of being predicted with reasonable accuracy.

Babylonian world map, early 5th century BC on a clay tablet, shows a flat, round world with Babylonia at the center. Eighteen of the animal constellations are named on the tablet.

Ancient Egypt

Ancient Egyptians paid little attention to observing and recording the movements of the Sun, Moon, and planets. Instead, special significance was attached to certain bright stars and constellations.

The ceiling of the Temple of Hathor at Denderah (above) depicts a magnificent series of ancient Egyptian constellations (illustrated, top).

Flood warning

By chance, the annual flooding of the River Nile— so vital to irrigating the crops that grew near the river —happened to coincide with the first sighting of Sirius, the sky's brightest star, as it rose in the east before dawn. It was deemed that the start of each new year would occur with the first new Moon following the reappearance of Sirius. When the system was adopted around 3,000 years ago, the rising of Sirius (known to the Egyptians as Sopdet) just before sunrise took place in early July; it now takes place around three weeks later, because the Earth's axis points to a slightly different place today than it did in those far off times.

Gods in the sky

A catalog of the heavens made in 1100 BC lists just five major constellations. In addition, 36 smaller star groups enabled the time to be calculated during the night—helpful tables allowing these calculations to be made have been found inscribed on a number of coffin lids. Osiris, the god of death, rebirth, and the afterlife, was represented by the bright, familiar constellation of Orion, and the swathe of the Milky Way represented the sky goddess Nut, who gave birth to the Sun god Ra each day. Circumpolar stars —those stars near the north celestial pole which

never set from Egyptian latitudes—were deities
known as the "Imperishable Ones."

Pyramid scheme

A collection of large stone slabs recently discovered
in Nabta in Egypt (in the Sahara Desert) represent the
world's earliest known astronomically-aligned
construction. Dated between 6,000 to 6,500 years
old, it is thought that the megaliths were erected by
the direct ancestors of the builders of the pyramids,
in what was once a fertile area. Climate change
caused the residents of Nabta to move eastward into
the Nile Valley, where the great pyramids at Giza
were built some time between 2700 to 2500 BC.
Astronomy was used to precisely position the
pyramids, since their sides are aligned almost exactly
north-south and east-west. The fact that they are
more accurately aligned east-west suggests that the
main positional sighting was made by observing the
rising and setting points of a star due east and west;
the star Acrab (Beta Scorpii) best fits for the era of the
pyramids' construction. It is likely that the positions
of the three great pyramids were meant to reflect the
belt stars of the constellation of Orion.

A belief that cosmic events and planetary
positions influenced human affairs is remarkably
absent throughout most of ancient Egyptian history.
Astrology did not figure until the Ptolemaic period,
from around the 3rd century BC, when the culture
was being influenced by nearby civilizations.
Perhaps the best star map depicting astrological
constellations can be found on the ceiling of the
Temple of Hathor (see left, page 30), constructed in
the first century BC.

**The three large pyramids at Giza
appear to mimic the position of
the three belt stars of Orion.**

Megaliths and medicine wheels

An awareness of the Universe, and a need to remain tuned in to its cycles, prompted ancient cultures in Europe and America to construct earthworks and stone buildings which were aligned with celestial events.

Little is known about the builders of these sites, but it is clear that they attached great importance to observing the Universe and perhaps being able to predict basic celestial events.

Hanging stones

Megalithic monuments dating back 5,000 years can be found across parts of western Europe and the British Isles. Weathered by the elements and damaged by people over the millennia, these battered gray stone circles and other prehistoric monuments make an eerie sight, but their former grandeur can be imagined. Perhaps the most famous of these ancient sites, Stonehenge, on the windswept Salisbury Plain in southern England, dates back to at least 2950 BC. Of unknown religious and ritual significance, Stonehenge and other megalithic constructions appear to have been used to make astronomical sightings of the Sun and Moon and to provide a means to calculate future solar and lunar solar events.

Medicine wheels

Many examples of the "medicine wheel"—a feature commonly constructed from stones laid on the ground which radiate from a raised central cairn to large circular stone rings—can be found across the United States and Canada. Smaller stone circles can often be found nearby these sites. Medicine wheels were constructed and used by native Americans from 4,000 years to a few hundred years ago. Astronomical alignments certainly exist in them, not only connected with the Sun and Moon but connected with the rising and setting points of numerous bright stars, which may have had special significance in native American religion and cosmology.

Ancient Stonehenge keeps a cold and silent vigil on England's Salisbury Plain.

Ancient China

Far removed from the cradle of western civilization, China developed its own form of astronomy, creating its own constellations and ideas about the workings of the Universe.

Skies under scrutiny

Over an almost continuous period spanning the 16th century BC to the end of the 19th century AD, court astronomers were appointed to observe and record changes in the heavens. This legacy of almost 3,500 years' worth of astronomy, in which sunspots, aurorae, comets, lunar, and solar eclipses and planetary conjunctions were noted, has provided us with a rich source of reference material.

Ancient Chinese astronomers created a catalog of stars visible with the unaided eye, divided the skies into constellations known as "palaces" and referred to the brightest star in each palace as its "emperor star," surrounded by less brilliant "princes." In the 4th century BC, the astronomer Shih-Shen catalogued 809 stars and recorded 122 individual constellations.

Instruments to aid naked eye observations were used extensively in ancient China, as they were in the west. In the 1st century, Lo-hsia-Hung constructed an armillary sphere—a device representing the celestial sphere, upon which were marked 365.25 divisions (for the days of the year), and rings for the celestial equator and the meridian. Lo-hsia-Hung's charming analogy for the Universe likened the Earth to the yolk within an eggshell, stating "the Earth moves constantly but people do not know it; they are as persons in a closed boat; when it proceeds they do not perceive it." In the 15th century an observatory was built on the southeastern corner of the city wall in ancient Beijing, which was equipped with a number of accurately calibrated sighting devices made out of bronze.

Cosmic firecracker

One of the most interesting records dates from 1054 AD and describes the sudden appearance of a "guest star" near the star we know as Zeta Tauri. This bright star, initially brilliant enough to be seen during the daytime and visible to the unaided eye for more than a year, was caused by a supernova—the catastrophic explosion of a massive star. Its remnants are visible today as the Crab Nebula.

Star signs

Astrology played an important role in ancient China. No fewer than 5,000 astrologers resided in 5th century Beijing! Twenty-eight constellations formed the ancient Chinese zodiac, through which the Sun, Moon, and planets progressed. Each of the five planets was designated its own element—Mercury, water; Venus, metal; Mars, fire; Jupiter, wood; Saturn, earth. A person's fate was supposedly determined by the relative position of the five planets, the Moon, Sun, and any comets that happened to be in the sky at the time of that person's birth.

Ancient Chinese astrological classification of comets—known as "broom stars." Each shape was said to foretell a different event.

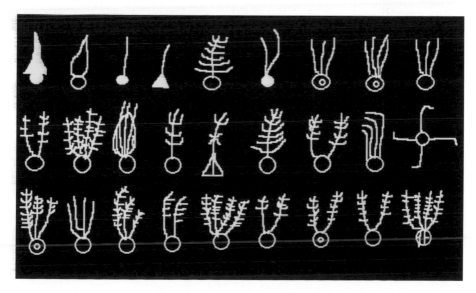

Ancient Greece

Many concepts about the Universe with which we are familiar today first arose in ancient Greece between 700 BC and 300 AD. Greek philosophers laid the foundations of modern astronomy.

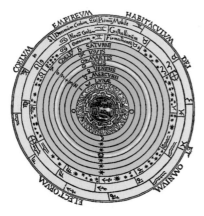

Eudoxus envisaged the Earth at the center of a series of crystal spheres upon which were fastened the Sun, Moon, individual planets, and stars.

Cosmic questioning

The Greeks benefited from the knowledge of the Universe that had been acquired by ancient Mesopotamian astronomers, and much of its ancient sky lore was also adopted and developed into fantastic celestial myths and legends. Quite unlike ancient Mesopotamia, conditions within the Greek civilization allowed scientific inquiry to flourish. Ancient Greek philosophers were the first to determine the size of the Earth, the distance of the Moon, and deduce the cause of solar and lunar eclipses. Greek philosophers inquired into the very nature of the cosmos, as attempts were made to explain objects and phenomena from the very small to the very big, from the basic atomic essence of matter, to the structure of the Universe.

Earth-centered Universe

In the 4th century BC, in accordance with Pythagoras' deduction that the circle and the sphere were perfect figures, Eudoxus devised a complete picture of the Universe, placing our planet at the center of a nest of 27 concentric transparent celestial spheres to which were attached the Sun, Moon, planets, and stars. Each of these Earth-centered spheres rotated around an axis shared with the Earth's axis. Using 55 spheres, the system was "improved" by Aristotle a century later. It was considered that the Sun, Moon, and planets were perfect spherical objects, and circular motions were deemed to be the only paths that could possibly be followed by celestial objects. This notion held sway throughout the entire era of ancient Greek astronomy and persisted through to the Renaissance.

In the 3rd century BC, Eratosthenes applied trigonometry to determine that the Earth is a sphere measuring around 25,000 miles in circumference, a remarkably accurate feat achieved by observing the length of the shadow cast by the Sun at noon from two widely separated places (Aswan and Alexandria) whose separation was known.

A contemporary of Eratosthenes, Aristarchus, used geometry to calculate the sizes and distances of the Moon and the Sun. By observing the Moon's half-phases and the angles made by the Earth, Sun, and Moon, Aristarchus concluded that the Sun was 19 times farther away than the Moon (the Sun's actual distance is 400 times that of the Moon). He observed the Moon passing through the Earth's shadow during lunar eclipses and concluded that the Moon was half the Earth's size.

Hipparchus, working in the 2nd century BC, made accurate measurements of the orbit, distance, and size of the Moon, and determined the distance of the Sun. Using an accurately constructed naked eye measuring device called an astrolabe, he observed and recorded the co-ordinates of around 850 stars to compile a star catalog. Another major star catalog was compiled in the 2nd century AD by Ptolemy, who included it in an encyclopedia of ancient Babylonian and Greek knowledge.

must know

Very little is known of Ptolemy's life. He made astronomical observations from Alexandria in Egypt during the years 127–41 AD. In fact, the first observation which we can date exactly was made by Ptolemy on March 26th, 127, while the last was made on February 2nd, 141.

Solar hub

Ptolemy's Earth-centered view of the Universe held sway in the west for around 1,500 years. Credit for the idea that the Sun, not the Earth, lies at the center of the Solar System is often given to Nicolaus Copernicus (1473–1543).

Copernican revolution

Heliocentric (Sun-centered) theories can actually be traced back to ancient Greece, to philosophers Pythagoras, Philolaus, and Aristarchus. However, in 1514 Copernicus was bold enough to produce a small handwritten pamphlet (but canny enough not to put his name to it) which challenged the very fundamentals of Ptolemy's geocentric (Earth-centered) view of the Universe. Copernicus stated that the stars are at an immense distance, compared to the distance from the Earth to the Sun. He was convinced that the Sun, not the Earth, lay near the center of the Universe, and that the apparent daily rotation of the heavens is caused by the Earth's rotation. Copernicus went on to explain that the apparent annual circuit of the Sun around the ecliptic is caused by the Earth revolving around the Sun, and the apparent retrograde motion of the planets is caused by the motion of the Earth along an orbit inside that of the outer planets.

His explanation of the phenomenon of retrograde motion, and dispensing with the need to introduce epicyclic planetary motions, is perhaps the most insightful and original of Copernicus' theories. Copernicus later laid out his potentially heretical heliocentric views of the Universe in his book *On the Revolutions of the Heavenly Bodies*, one of the first copies of which was given to him as he lay dying in 1543.

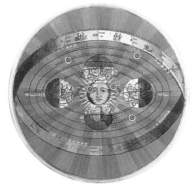

Placing the Sun at the center of the Solar System, Copernicus' heliocentric theory saw the Earth as one of six orbiting planets.

Copper-nosed enquirer

Tycho Brahe (1546–1601) is regarded as the last and greatest astronomer of the pre-telescopic era. A hot-headed Danish nobleman, Tycho lost part of his nose in a sword duel and later replaced it with a copper prosthesis. In an attempt to refute Copernicus' view of the Universe, Tycho began making precise measurements of the stars and the movements of the planets with the aid of quadrants and cross-staffs from his observatory at Uraniborg on the island of Hven. Tycho's careful naked-eye observations went on to provide plenty of evidence disproving the old established notions of the Earth-centered Universe.

Tycho Brahe, a great pre-telescopic astronomer. He used naked eye devices to make incredibly accurate observations of planetary motions.

Understanding the Solar System

In the early 17th century, a potent combination of theory and observation expanded human notions about the Universe. The telescope opened everyone's eyes to the cosmos.

Kepler, planetary law-giver

Despite finding evidence to support the idea that the Earth was a planet in orbit around the Sun, Tycho Brahe remained highly sceptical of the idea. He clung to the old notion of an Earth-centered Universe whose motions would eventually be explained as soon as the right mathematical model was found. Tycho's assistant, Johannes Kepler (1571–1630), had none of his teacher's confidence in the geocentric theory, and used Tycho's extensive observations to place the heliocentric theory on a firm scientific footing. Profoundly religious, Kepler was convinced that God had created the Universe in accordance with mathematical rules, and that a knowledge of these rules was within human comprehension.

Galileo's telescopic observation of the Pleiades. Only a handful of stars in this cluster can be seen with the unaided eye.

Telescopic revelations

Having heard about a newly invented optical
instrument that made distant objects appear larger,
Galileo Galilei, a professor of mathematics at Padua

discovered four small satellites orbiting Jupiter,
which are still referred to as the Galilean Moons.
Venus displayed phases, proving that it was a globe
in orbit around the Sun. Sunspots proved that our
central star is by no means perfect, and the Moon's
surface was revealed as a world with dark plains,
craters, and high mountains. Delving into deep
space, Galileo observed that the glowing band of the
Milky Way was made up of multitudes of faint stars.

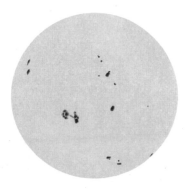

**An observation of the Sun made
by Galileo in July 1613 shows detail
within sunspot groups.**

The orbs around us

During the 17th century, as telescopes improved, an increasing number of scientists and amateur astronomers were keen to learn more about the Universe by scrutinizing celestial realms previously hidden from human view.

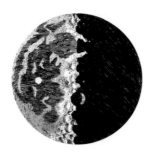

The Moon and its craters, from an observation made by Galileo in November 1609.

Reality bites

People began to question the workings of the Universe, and they now had optical tools with which to study the heavens more closely. Yet, despite the evidence, there were many who did not accept the reality of what the newly invented telescope revealed about the Universe. Some had seen the power of the telescope with their own eyes but believed that it was trickery or the work of the devil. Galileo and others who supported the Copernican view that the Earth was a planet in orbit around the Sun were denounced as heretics. Even Galileo succumbed to pressure when he was asked by the Inquisition to refute the idea.

Our lunar companion

The Moon is so big and bright that plenty of detail can be seen on its surface through even the most basic optical equipment—a fact that has delighted lunar observers from Galileo to the present day. It is not surprising that many early telescopic observers chose to study the Moon, to draw its features, and to map its surface. The Moon's landscape really did resemble parts of the Earth—Galileo had likened it to parts of Bohemia—so did it have an atmosphere and could it support life? Large dark patches visible with the unaided eye were discovered to be relatively flat gray plains. These areas became known as "maria" (Latin: seas), but it was plain to see through the telescope eyepiece that they did not represent bodies of water.

The mid-17th century saw the publication of a number of detailed lunar maps. One by Johannes Hewelke (Hevelius) was published in his *Selenographia*, complete with names for lunar features based upon geographical landmarks, like Sicily, Mount Etna, and the Mediterranean Sea. From his private observatory, Hevelius also made accurate measurements of star positions and produced the *Uranographica*, the most advanced star atlas of its time; the names for his seven new constellations are still used by astronomers today.

At around the same time, Giovanni Riccioli published an accurate lunar map which incorporated nomenclature that is still current, including his names for the Moon's seas, such as Mare Tranquillitatis (the Sea of Tranquillity), and many of the larger craters like Copernicus (Mount Etna on Hevelius' map)—familiar names to modern watchers of the Moon.

Giovanni Riccioli's Moon map of 1651.

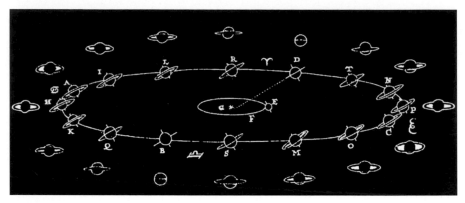

Christiaan Huygens' explanation of Saturn's rings and their changing appearance.

Planetary perceptions

Fresh telescopic revelations answered some age-old questions about our immediate planetary neighborhood but raised many more new ones.

Aerial adventures

In the Netherlands, Christiaan Huygens constructed lengthy refracting telescopes whose lenses were suspended on frames in the air and moved by means of ropes and pulleys. Such unwieldy devices—some of them measuring more than 26oft (80m) from objective lens to eyepiece—were required to overcome a defect inherent in simple optics known as chromatic aberration. Caused by the inability of a single objective lens to focus all the colors within white light to a single point, chromatic aberration causes bright objects to appear surrounded by colored fringes. The longer the objective lens, the less evident the degree of aberration. Huygens also improved eyepiece design by introducing the two-lensed Huygenian ocular—a design still provided today with many budget telescopes. Using these ungainly aerial telescopes he discovered Jupiter's equatorial bulge—the consequence of Jupiter's rapid spin and its gaseous composition. Bright polar caps were discovered on Mars, in addition to a dark V-shaped feature known as Syrtis Major. Almost half a century after Galileo had discovered the Solar System's first planetary satellites, Huygens discovered Saturn's largest satellite, Titan, in March 1655. A conundrum which had baffled astronomers since Galileo—mysterious appendages that seemed to cling to Saturn's side and vary in size over the years—was finally solved by Huygens, who explained that Saturn had a flat ring system that nowhere touched the planet.

Under Paris skies

In the late 17th century Giovanni Cassini helped establish the Paris Observatory as the world's first fully equipped national observatory. Cassini measured the shape of Jupiter and its rotation by making timings of the Great Red Spot as it transited the planet's central meridian, a technique still used today. He timed Mars' rotation, finding it to be just 37 minutes longer than the Earth's. Observations of Mars' position were also used to determine the scale of the Solar System. Cassini correctly suggested that the rings of Saturn consist of countless tiny moonlets, all in independent orbits in the same plane. He discovered a gap in the rings of Saturn—referred to as Cassini's Division—and found four more Saturnian satellites. Cassini discovered a delay between predictions of Jupiter's satellite phenomena and timed observations; this enabled his colleague, Ole Roemer, to determine the speed of light fairly accurately.

The speed of light was determined in 1675 by noting the delay between predicted and observed timings of Jovian satellite phenomena. Predictions which correctly matched observations at Earth 1 (opposition) were found to be around 20 minutes early six months later at Earth 2, because Jupiter's light had to travel farther to the observer.

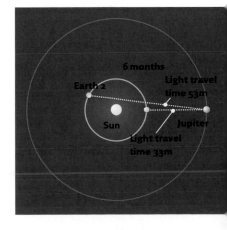

The orbs around us | 45

must know

Halley and the comets
Newton's friend, Edmond Halley, tackled the problem of cometary orbits. He reasoned that these occasional visitors to the inner Solar System were influenced by the same gravitational laws as the planets, and that they moved in highly elliptical paths around the Sun. He calculated that comets observed in 1531, 1607, and 1682 were one and the same object in an orbit that brought it close to the Sun every 76 years. Halley correctly predicted the comet's appearance in 1758. Sadly, he died several years before the comet, which now bears his name, came into view.

Newton's Universe

Isaac Newton made a tremendous impact on our view of the way the Universe worked. His laws of motion and universal gravitation, formulated in the late 17th century, allowed astronomers to precisely predict the movements of objects, from an apple falling from a tree to the motions of the Moon, planets, and the stars themselves.

Newton invented a new form of telescope—the Newtonian reflector—which used a precisely shaped concave mirror to collect and focus light. Not only did this overcome the problem of false colors but allowed larger telescopes to be constructed more easily, since only one surface required shaping, rather than two surfaces of an optically clear piece of glass for the objective lens of a refracting telescope. This type of instrument became very popular among professional and amateur astronomers, and remains so to this day. Newton's hand-made original has a mirror just 2in (5cm) across.

Ikeya-Zhang, a periodic comet, observed by Peter Grego in May 2002.

The active nucleus of Halley's Comet, imaged by ESA's Giotto spaceprobe in 1986.

Celestial mechanics

Originally invented by clockmaker George Graham in 1710, finely crafted clockwork models of the Solar System, known as orreries, became popular in the early 18th century. They show how society had come to confidently embrace the heliocentric view of the Universe, which only a century before had been deemed heretical. Some of the best orreries showed all the planets in the Solar System, and their known satellites (though not to scale), which all revolved along their respective orbits at their correct relative speeds.

Planetary perceptions

Astronomers of the golden era of observation managed to ascertain a great deal about the Solar System by peering through their telescope eyepieces.

Golden age of observation

For three centuries following Galileo, most astronomy was carried out with eyes at the telescope eyepiece, and records of the Sun, Moon, planets, and objects farther out in the Universe were made by hand, either by writing down information or by making drawings on a sketchpad. There was a great mix of amateur and professional contributions to our knowledge of the Universe. Advances in telescope optics, including the invention of objective lenses that eliminate much of the false color produced by refracting telescopes, allowed for bigger and better instruments with which to probe the Universe.

Mercury, shown in a patchwork of images returned by Mariner 10 in 1974—the only probe ever to have visited the Sun's nearest planet.

Cook-chilled planet

Mercury, the planet nearest the Sun, displayed phases, just like its neighbor Venus. Features on Mercury—a tiny planet and difficult to observe because of its proximity to the Sun—were vague, but it was clear that Mercury had little or no atmosphere. Some astronomers thought that Mercury was tidally locked by the Sun's gravity, so that one face was turned toward the Sun, while the other experienced perpetual night. It was thought that the day side of Mercury was the hottest place in the Solar System, while its night side was incredibly chilly.

Cloud-swathed Venus

Venus was found to be an earth-sized world. It
showed few features, and those which did appear
drifted and faded over a period of a few days. Since
the planet's surface was permanently hidden from
view, speculation was rife about the conditions on
Venus' surface. Some astronomers imagined that
Venus was in a phase similar to Earth's Carboniferous
Period, with lush jungles and forests. Observers had
frequently noted that the cusps of Venus appeared
dazzlingly bright, a phenomenon that some
astronomers speculated was due to brilliant, highly
reflective icy polar caps at the planet's poles. Other
observers had noted points of light on the planet's
unilluminated portion, near the planet's terminator
(the line separating the night and day sides of a
planet), and speculated that these were the summits
of incredibly lofty mountains peeking above the
Venusian atmosphere.

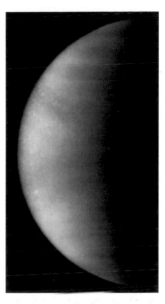

Venus—mysterious and very
cloudy.

Quiet Moon

Our own Moon displayed no obvious changes, other
than the shadowplay of topographic features during
the lunar surface's illumination by the Sun for two
weeks during each lunar month. The Moon clearly
had no atmosphere; no winds whipped across its gray
surface, no clouds scuttled across its skies, and rain
never quenched the dry Moon dust. From time to
time astronomers reported (and continue to report),
strange activity on the Moon, including colored glows,
flashes, and obscurations of surface features, the
causes of which may include meteorite impacts and
gaseous emissions. Nonetheless, the Moon is one of
the most inactive worlds in the Solar System.

Strange red world

Mars was (and still is) perhaps the most fascinating of all the
planets. It can only be observed for a few months at intervals
of every two years or so, when the Earth is near enough to Mars
to observe its features in any kind of detail. Icy polar caps,
which grow and shrink with the seasons, crown the Red Planet,
and bright clouds sometimes develop in its thin atmosphere.
The Martian surface is crossed with light and dark tracts.
Unlike the Moon's maria, the markings on Mars vary in intensity
and outline from season to season and from one apparition to
the next, while retaining their general form and position in the
long term; all of which makes it an intriguing world.
Astronomers reasoned that Mars was the most Earth-like of
all the planets, and perhaps the nearest place in the Universe
to harbor some form of life.

Mars: an abode of life?

In 1877, Giovanni Schiaparelli claimed to have observed linear
markings stretching for thousands of miles across Mars.
Believing these features likely to be natural geological
formations, he termed them "canali" (Italian: "channels").
Schiaparelli, a respected astronomer, was careful not to
speculate that the canali might in some way be linked with life
on Mars. The observations fueled to a longstanding debate on
the possibility that life may exist on Mars. Soon, other
astronomers claimed to be able to trace the Martian "canals."
Some proclaimed them to be vast waterways hewn out of Mars'
deserts by an advanced Martian civilization. Percival Lowell, the
most vociferous advocate of this theory, wrote several books on
the subject in which he claimed that the lines on Mars were
tracts of vegetation bordering canals distributing meltwater
from the polar icecaps. Sadly, intelligent Martians have never
existed, and the canals of Mars were caused largely by
psychological or physiological illusions.

Mars, observed by Christiaan Huygens in 1659. The dusky V-shaped tract of Syrtis Major is depicted.

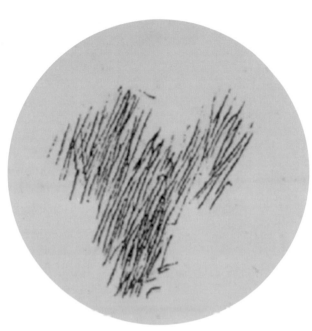

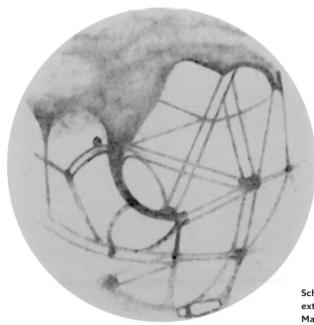

Schiaparelli recorded an extensive network of lines on Mars in this observation of 1888.

Celestial vermin

From the Earth, none of the many thousands of known asteroids are bright enough to be seen with the unaided eye, so it isn't surprising that nearly two centuries had elapsed after the invention of the telescope before the first asteroid was discovered.

Police squad

On January 1st, 1801—the first day of the 19th century—Giuseppe Piazzi discovered the first asteroid, which he later named Ceres. Ceres was far too small for its disk to be made out at the telescope eyepiece, but the methodical astronomer charted the object's slow movement among the stars of Taurus. Piazzi thought that the object might be an incoming comet, but after a while it became clear that the new object was a minor planet orbiting the Sun between Mars and Jupiter. Reasoning that there might be yet more minor planets to be discovered, a group of European astronomers organized themselves into the "Celestial Police" to hunt them down. Minor planet Pallas was discovered in 1802, Vesta in 1804, and Juno was found in 1807.

Asteroid Gaspra—a 12 mile- (19km-) long potato-shaped chunk of cratered rock.

By the end of the 19th century, several hundred minor planets were known. During the 20th century, so many asteroids were being discovered photographically that they were sometimes rather unkindly referred to as "the vermin of the skies." Today, the orbits of more than 277,000 asteroids are known.

Finding faint fuzzies

From the 18th century onward, comet-hunting became highly competitive. Having discovered a comet through a telescope perhaps many weeks before it was bright enough to be seen with the naked eye, its finder would be rewarded by having his or her name attached to it. During his comet searches, Charles Messier sometimes mistook deep sky objects—dim nebulous patches far beyond the Solar System—for comets. Messier sketched and noted the positions of around a hundred of these deep sky objects, publishing them in his famous Messier catalog of 1784. Nicknamed "the comet ferret," Messier went on to discover 20 comets, and his list is still used by astronomers eager to observe the northern hemisphere's brightest deep sky objects.

Part of the path of the Great Comet of 1769, one of Charles Messier's 20 comet discoveries.

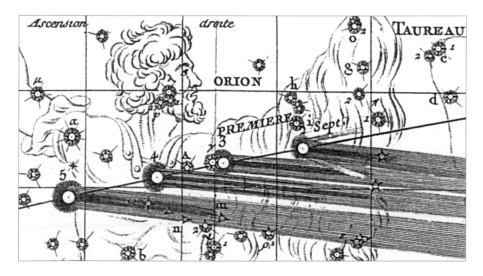

Jovian giant

Astronomers marveled at Jupiter, the largest planet; with its four Galilean moons, it appeared like a miniature Solar System. Jupiter's polar flattening and a series of bright and dark bands running parallel to the planet's equator had been noted by Cassini at the Paris Observatory in the 1670s, who also timed the appearance of certain dark features on the planet's disk and calculated that Jupiter turned on its axis once every 9 hours and 55 minutes. Robert Hooke made an interesting series of observations a decade before Cassini, in which he may have been the first to observe Jupiter's Great Red Spot, a long-lived Jovian weather system which can still be seen to churn up the

Jupiter and its Great Red Spot. Jupiter's big moon Io lies to the lower left.

planet's atmosphere today. For some reason, astronomers neglected detailed scientific studies of Jupiter until the latter half of the 19th century.

Ringed planet

As we have seen, the nature of the curious appendages of Saturn had baffled astronomers since Galileo, but they were finally explained by Huygens as being a vast, flat ring system. Cassini discovered the main gap in the rings of Saturn, and more structure within the rings was revealed in the following century, including a smaller gap in the outer rings, known as the Encke division, and a ghostly inner ring which became known as the Crepe Ring. Like Jupiter, Saturn is a gas giant, and its rapid spin causes a pronounced equatorial bulge. Saturn's atmosphere is, however, far less dynamic than that of Jupiter, and astronomers became excited at any opportunity to see distinct detail, including large white ovals which brew up every few decades.

Expanding Solar System

Using home-made Newtonian reflecting telescopes, the gifted William Herschel established himself as the world's most prolific astronomer. In 1781 he doubled the scale of the known Solar System when he discovered the planet Uranus.

Observations of Uranus' position in the decades following its discovery suggested that its orbital path was being affected by the gravitational pull of a large unknown planet lying further out in space. Working on this premise, Urbain Leverrier calculated the probable position of this mystery planet. Using Leverrier's prediction, Johann Galle discovered the new planet—later named Neptune—at the Berlin Observatory in 1846. A similar search based on a mathematical prediction enabled Pluto, the most distant object "officially" designated as a planet, to be discovered on photographic plates taken at the Lowell Observatory by Clyde Tombaugh in 1930.

Saturn, in apparent close proximity to the Crab Nebula, observed in January 2003. Actually, the unique positioning of these two astronomical features in the same part of the sky allowed this photograph. Saturn is around 70 light minutes away, while the nebula is 6,500 light years distant.

Birth of the Solar System

With this knowledge of the astonishing range of worlds orbiting the Sun that grew more intimate with each discovery—and a sure realization of the vast scale of the Solar System—astronomers began to make attempts to explain how it all came into being.

Clouded visions

In the late 18th century, Immanuel Kant and Simone Laplace independently arrived at the conclusion that the Solar System was formed from a vast cloud of dust and gas known as a nebula. Collapsing under its own gravity, the original solar nebula began to spin ever faster—just as a ballerina spins faster by pulling in her arms while performing a pirouette—and the nebula became flattened like a disk. Over time, as the massive core of the nebula condensed to form an ever hotter embryonic Sun, rings of material were thrown out. Disturbances within these rings condensed into planets, and a similar process around them formed the planetary satellites. It was suggested that many nebulae were the birthplaces of more distant Solar Systems.

Passing star

Planets condensed within rings around a contracting embryonic Sun, according to the nebular theory.

At the beginning of the 20th century, Thomas Chamberlin and Forest Moulton came up with an intriguing theory, arguing that a massive star passed so close to the Sun that its gravity drew out a huge mass of material from the solar surface. After

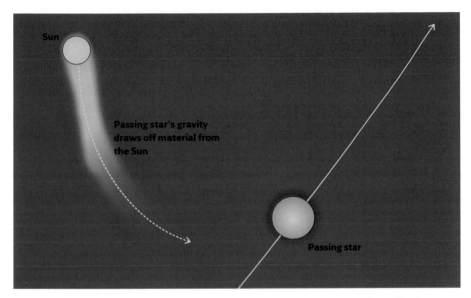

cooling and condensing into solid matter, this debris ultimately
coalesced to form the planets, and remnants of this material
formed the Asteroid Belt.

Interstellar origins

Our modern view of the formation of the Solar System begins
5 billion years ago, with a gravitational ripple within a cloud of
cold interstellar gas and dust. One of the denser parts of the
cloud collapsed under its own gravity, and the nebula began to
spin, producing a disk of material. At its center, increasing
pressure and temperature caused it to grow so hot that
thermonuclear reactions were triggered, and the Sun was born.
The solar wind blew away most of the nebula's dust and gas,
leaving behind just the larger, denser clumps of matter which
had formed in the disk. These were to become the Sun's planets
and their moons, asteroids, and comets. Some of this material,
virtually unchanged since the Solar System's formation, is
thought to lie within icy cometary nuclei.

According to one
long-abandoned
theory of the Solar
System's formation,
a passing star drew
off a mass of
material from the
Sun, which later
condensed into
planets.

A bigger picture

Although the idea that the Sun might be just an ordinary star around which the Earth orbited had originated in ancient Greece, it was not until the 18th century that observations began to support this notion.

Home galaxy

Galileo's tiny telescope showed that the Milky Way was composed of multitudes of stars that were too faint to be resolved individually with the unaided eye. If the stars themselves were like the Sun, but so far away that they appeared as mere points of light, then perhaps the Sun itself didn't lie at the hub of the cosmos, but was one of a broader mass of stars contained within a vast star system—a galaxy containing countless thousands of stars. In the mid-18th century, Thomas Wright made the perceptive suggestion that the band of the Milky Way was caused by our view from deep within a vast, flat millwheel-shaped star system; he found no evidence that the Sun lay at the galaxy's center or that the system was finite in breadth.

Herschel's insights

From southern England, using his own home-made telescopes, William Herschel conducted a thorough "review of the heavens"

Thomas Wright's 1750 concept of the galaxy's layout.

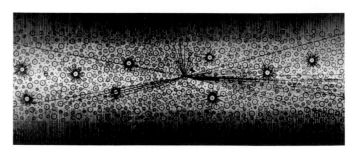

during the late 18th century. He succeeded in producing a catalog of star clusters and nebulae that surpassed the relatively modest work of Charles Messier. Herschel noted the relative motion of stars, proving that they were not simply points of light fixed on some distant heavenly vault. Some stars were shown to be physically associated with one another, having common orbits around each other. Herschel understood that if the Sun really was moving through a galaxy full of stars, then those stars in the direction of the Sun's motion would appear to be moving apart, while those "behind" the Sun would appear to be moving toward each other. Measurements of star positions enabled him to discover the Sun's motion through space.

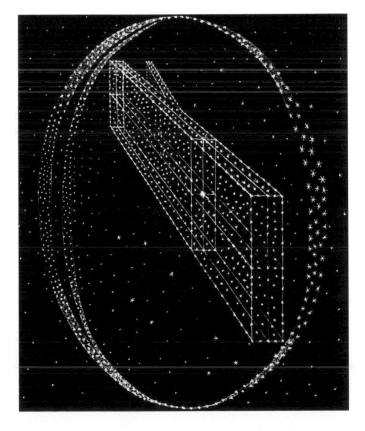

William Herschel's 1784 speculation on the Milky Way's structure, showing (center) a representation of the layout of the galaxy in the vicinity of the Sun and (outside) how the galaxy might appear from a great distance.

Charting the stars

Look at any antique star map and it can be hard to distinguish the stars from the beautiful depictions of mythical constellation figures. They may look like quaint works of art, but the star maps produced by astronomers like Johann Hewelcke in the mid-17th century and John Beavis in the mid-18th century were depicted with the best accuracy then attainable, each star's position having been carefully noted at the telescope. Accurate charts enabled astronomers to identify new planets, novae, and deep sky objects, and it became possible to accurately measure the motion of the Moon and planets.

Aberration of light

In 1728, James Bradley discovered that the observed position of the stars varies slightly as the Earth moves around the Sun—every star appears to move in a tiny self-centered ellipse each year, the star's latitude determining the flatness of its ellipse. This phenomenon is known as the aberration of light. It occurs

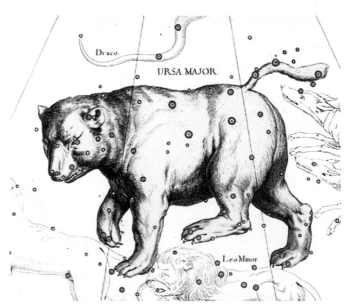

Johann Hewelcke's 1690 chart of Ursa Major, the Great Bear.

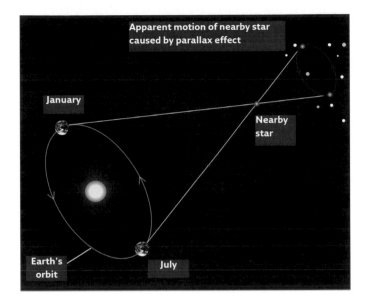

Apparent motion of nearby star caused by parallax effect

January

Nearby star

Earth's orbit

July

Illustration of the apparent motion of a nearby star against the distant celestial background, caused by our changing line of sight. The effect is greatly exaggerated for clarity.

because the Earth, travelling at a speed of 37 miles per second in its orbit around the Sun, intercepts starlight travelling at 186,000 miles per second. It's just like the effect caused by moving through a rain shower. Even though the rain is falling straight down, anyone moving through the shower must tilt their umbrella forward, since the raindrops appear to be approaching from an angle.

Plumbing cosmic depths

Following up on Bradley's star charting work, Friedrich Bessel was the first to observe the regular change of a star's position due to the changing line of sight as the Earth orbits the Sun, an effect known as stellar parallax. Stars nearer to the Earth show the greatest parallax, while the parallax of increasingly distant stars is more difficult to measure. Bessel's measurement of the parallax of the star 61 Cygni allowed its distance to be calculated —the first measurement of the distance to the stars. It was found that 61 Cygni was an incredible 11 light years away.

Observations of Cepheid variable stars allow astronomers to gauge the scale of the Milky Way and the distance of globular clusters, the Magellanic Clouds, and more distant galaxies.

Small Galaxy, big Universe

Comfortable notions that the entire Universe was dominated by the Milky Way began to be challenged during the late 19th century. Photography began to reveal more of deep space than could be seen by the visual observer, and the new science of spectroscopy laid bare the composition of stars and nebulae.

Island Universes

Even the keenest visual observer using a large telescope will struggle to discern detail within most of the nebulae that dot the heavens. Using the "Leviathan" at Birr Castle in Ireland—a telescope with a huge 6ft (1.8m) diameter mirror, which was then the largest telescope in the world—William Parsons visually observed nebulae, hoping to discern individual stars within them. In 1845 he discovered the spiral nature of one particularly bright nebula, now known as the Whirlpool Galaxy. Other nebulae were noted to have a spiral form, too—some of these, like the Whirlpool, appeared face-on, while others were presented at a variety of angles, including some that were almost edgewise.

Deep sky diversity

When photography began to be used extensively for deep sky studies from the late 19th century onward, the true grandeur of many deep sky objects was uncovered for the first time. Deep sky objects were divided into a number of distinct groups—open star clusters, globular star clusters, amorphous gaseous nebulae, planetary nebulae, and spiral nebulae. Astronomers began to question whether the nebulae really did represent stages in the evolution of distant solar systems—especially that spiral nebulae had been shown to be very large objects packed with faint stars and at a great distance from the Earth.

Cosmic yardstick

In 1908 Henrietta Leavitt carefully studied photographs of the Milky Way's near neighbor, the Small Magellanic Cloud. She identified a number of stars whose luminosity varied regularly. The period of time over which they varied from minimum to maximum brightness appeared to depend on their actual luminosity. Leavitt was able to state this with reasonable certainty, since she assumed that the stars within the Small Magellanic Cloud were all about the same distance from us.

The Whirlpool Galaxy, sketched in 1845 by William Parsons at the eyepiece of the 6ft (1.8m) 'Leviathan' at Birr Castle, Ireland.

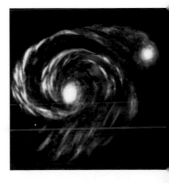

A number of these stars, known as Cepheid variables, were identified in our own Milky Way. By knowing the actual brightness of a star its true distance could be measured. With this useful tool the vast distances between stars, star clusters, and galaxies could be known. After studying Cepheids in globular clusters around the Milky Way, Harlow Shapley produced the first map of our galaxy. Shapley's Milky Way was a vast system of a hundred billion stars, arranged in a flattened disk; the Solar System appeared to be located some distance from the galactic center.

Edwin Hubble (1889-1953): pipe-smoking genius, Illinois state high jump champion, and discoverer of the expanding Universe.

Hubble's Universe

In the 1920s Edwin Hubble used the 8ft (2.5m) telescope at Mount Wilson in California to search for Cepheid variable stars within a number of spiral nebulae. Questions about their size and distance had been asked for decades, and Hubble finally provided measurements which astounded astronomers. Far from being residents of our own Milky Way, objects like the spiral nebulae in Andromeda and Triangulum were shown to be independent galaxies of comparable size to the Milky Way, each made up of thousands of millions of stars and located at least 25 times farther than the most distant stars of our own galaxy. It was a sobering thought that the light from these galaxies had set off in prehistoric times, millions of years ago.

Redshifted cosmos

By studying the light from distant galaxies, Hubble went on to conclude that the Universe is expanding, and that the further away a galaxy is, the faster it appears to be moving. To understand how Hubble arrived at this remarkable conclusion, let's take a look at a useful tool which astronomers use, the spectroscope. Spectroscopes use a prism to split light into its component colors. Narrow dark bands called Fraunhofer lines can be seen within the resulting rainbow of light—these are produced by certain chemical elements within the stars. The Fraunhofer lines in the spectrum of an object moving away from us at a very high velocity will appear to be shifted towards the red end of the spectrum; this is called redshift. Hubble showed that the further a galaxy is from us, the greater its redshift—in fact, galactic redshifts increase in proportion to their distance from us. It was later shown that galaxies are distributed throughout space in clusters and superclusters. Our own galaxy was revealed to be just one of billions of galaxies spread throughout space and time.

Wavelengths of light from approaching objects are compressed and shifted toward the blue end of the spectrum, while light from objects receding from the observer is stretched out into the red.

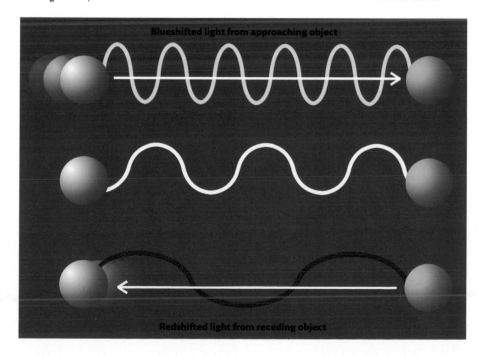

Blueshifted light from approaching object

Redshifted light from receding object

Expanding Universe

Of course, the observation that all galaxies are moving away from us is simply caused by our cosmic perspective within an expanding Universe; the same observation could actually be made from any point within the Universe. Imagine a semi-inflated balloon upon which are drawn a number of dots; these dots represent galaxies. As the balloon (representing the fabric of space-time) is inflated further, and its surface expands, the dots all move away from each other—there's no central point of expansion on the spherical surface of the balloon. In the same way, there's no central point in this expanding Universe of ours, away from which everything is moving.

The Big Bang theory

Scientists have observed that if the Universe is expanding, then there must have been a time in the remote past when everything in the Universe was contained within a single point. This point—the primordial atom—exploded to create time, space, and

Fluctuations in the cosmic microwave background radiation show the seeds of galaxy formation in the early Universe.

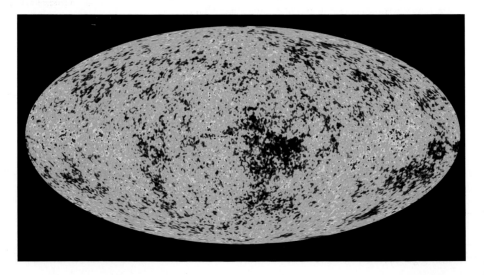

the entire Universe. Cosmologists are now fairly confident that this event, known as the Big Bang, took place between 13 and 14 billion years ago.

Blast from the past

Good evidence supporting the Big Bang theory comes in the form of an echo from the beginning of time—the cosmic microwave background (CMB) radiation. In 1965 radio astronomers Arno Penzias and Robert Wilson detected a diffuse background radio signal which appeared to emanate uniformly from all directions. Puzzled by the signal, Penzias and Wilson eliminated all possible mechanical and instrumental causes—including pigeon droppings in the antenna—yet the signal stubbornly remained. They eventually concluded that it was nothing less than a relic of the hot primordial fireball, dating to around 100,000 years after the Big Bang. Originally a temperature of 3,000°K (Kelvin) or 2,727°C, the expansion of the Universe has redshifted the radiation into microwave wavelengths, where it permeates the cosmos at a temperature of just 3°K above absolute zero.

Ripples in the cosmic fabric

Twenty-five years after the discovery of the CMB, the Cosmic Background Explorer satellite detected small variations in the background radiation intensity across the sky. Since confirmed by more sensitive instruments, and mapped out in detail by the Wilkinson Microwave Anisotropy Satellite (WMAP) in 2003, these fluctuations in the temperature of the CMB represent large-scale ripples in the once-smooth fabric of space. It is not understood what caused these fluctuations, but it is thought that the first physical structures in the Universe —young galaxies and galactic clusters—grew from them.

want to know more?

Take it to the next level...
▶ **More on the Solar System** 86–120
▶ **Delve into our galaxy** 122–51
▶ **Cepheid variables, cosmic yardsticks** 134
▶ **Radio astronomy** 178

Other sources...
▶ **Join the Society for the History of Astronomy.**
▶ **There are many great books available on basic astronomy (see Further Reading, p188). Many excellent books clearly introduce the wider Universe and cosmology without delving into complex maths!**

Weblinks...
▶ **The Society for the History of Astronomy. UK-based, but membership is open to all www.shastro.org.uk**
▶ **A site for those curious about the history of astronomy curious.astro.cornell.edu/history.php**
▶ **Cosmology 101 – a guide to the ever-expanding Universe map.gsfc.nasa.gov/m_uni.html**

3 Third rock

Planet Earth is the only place in the Universe
known to have developed life. Humans
have developed the capacity to learn about
its immediate environment and have a need
to understand its place in the wider Universe.
Although a tiny step compared to the infinite
distances in the firmament, humans have
also traveled to its closest neighbor, its
satellite the Moon.

Our cosmic foothold

We live on an interesting planet at an interesting time in its long history. This spherical rock was once considered vast beyond comprehension. The view from space has shown us that our home planet is finite and all too fragile.

Planet Earth

Planet Earth, the fifth largest planet in the Solar System, measures 8,000 miles (12,756km) across at the equator. A spacecraft in low Earth orbit completes a circuit around the globe in just 90 minutes; peering through their porthole, astronauts aboard the International Space Station see the landmass of North America slip by, from the mountains of the west coast to the plains of the east, in less than eight minutes.

Spinning on its axis once every 24 hours, the Earth orbits the Sun at an average distance of 93 million miles (150 million km), taking just over a year to make each near-circular circuit at an average velocity of more than 66,000 miles (107,000km) per hour. Held to the Earth by its gravitational bonds, and sharing the Earth's velocity, we can feel none of this directly. Gravity also holds in place the Earth's atmosphere—a mixture of 78 per cent nitrogen, 21 per cent oxygen, and a few other constituents.

Natural beauty

With its swirling white cloud systems, the Earth has often been called the "blue marble." Images from space have given us a different perspective of the beauty of the world—perhaps our raised awareness of the Earth's fragile ecology has stemmed in part from the impact of photographs of the Earth taken by NASA astronauts. Earth's broad blue oceans cover around three quarters of its surface, and its poles are capped by brilliant white icecaps. Land masses appear in a variety of colors, from yellow-brown to verdant green, and astronauts have often been struck by the fact that there appear to be no obvious political or national boundaries on the Earth's surface. Upon closer scrutiny, astronauts find that the generally gray cities around the globe are fairly easy to discern, along with the grid patterns of city streets. At night the globe appears illuminated with a glittering spider's web of artificial lighting. Additional nocturnal lighting comes from forest fires and volcanic eruptions, while high in the atmosphere, auroral displays dance around the geomagnetic poles, and meteors strike their burning paths.

This view of Earth rising above the Moon was captured by Apollo 11.

Terra, formed

Around 4.5 billion years ago, the newborn Sun was surrounded by an orbiting disk of gas and dust. Over time, gravitational attraction between particles of matter in this disk caused larger clumps to gather, and in turn these snowballed to form individual planets.

Planetary foundry

It may now look serene, but the blue planet had a hot, violent time during its youth.

One of these embryonic worlds became the Earth. As the planet accumulated more material, impacts became more violent, causing heating which gradually raised the planet's temperature. With its increasing mass, the interior began to

compress under its own gravity, producing more heat within the young Earth. Eventually, temperatures within the Earth became high enough to melt its iron and nickel constituents. As these heavier metals sank towards the Earth's center, forming the planet's core, lighter material such as silicon, magnesium, and aluminium rose to form the planet's mantle and crust.

Conditions on the early terrestrial crust were hardly stable; meteorites, asteroids, and comets continued to smash into the planet, while molten material burst through weak points in the crust to spread across the surface as lava flows. The process continued relentlessly for more than a billion years, until most of the debris within the inner Solar System had been swept up by the planets, impact rates dwindled, and the Earth's rocky crust began to thicken and consolidate.

Atmosphere, oceans, and life
Continued volcanic eruptions fueled by the Earth's searingly hot interior provided gases such as hydrogen, nitrogen, carbon monoxide, carbon dioxide, and water vapor, forming a primitive atmosphere that was far from able to sustain human life. Initially, any water vapor condensing in the atmosphere and falling as rain simply boiled off the Earth's surface, but it eventually cooled sufficiently for liquid water to exist. Seas began to form, filling in the lowlands and ancient impact craters, and from time to time comets—large iceballs left over from the Solar System's birth—also added to the Earth's complement of water and other material.

Around three billion years ago, primitive lifeforms, capable of converting the abundant carbon dioxide and water in its environment into oxygen, developed on this wild young world. Oxygen-dependent organisms followed, which further transformed the atmosphere; the Earth's ecosystem has since grown increasingly complex, accompanied by interdependence between flora, fauna, and individual species.

Jigsaw planet

Today, everyone is familiar with the way the Earth's surface is laid out, with its large continents separated by vast oceans. Timeless though the Earth's face may appear, our view of its true permanence underwent a change of seismic proportions during the 20th century.

Get the drift

Around a century ago, Alfred Wegener put forward the idea that the Earth's continents were slowly moving with respect to each other over a very long period of time. His startling theory of continental drift was bolstered by a good deal of evidence: as anyone armed with a map of the world can see, the eastern coastline of South America and the western coastline of Africa would make a pretty snug fit if they were positioned next to each other. More compelling still, the ancient rocks found in Africa, their fossil record and glacial markings matched those of South America exceedingly well. Wegener's theory of continental drift was initially rejected as being utterly fanciful; one geologist called it "utter, damned rot!." But widespread acceptance of continental drift came in the 1960s, when seafloor spreading was discovered.

When continents divide

It is thought that all the continents were once joined in a single supercontinent known as Pangaea, surrounded by the oceanic tract of Panthalassa. Around 200 million years ago, the radioactive decay of elements within the Earth's core had provided enough heat in the mantle to drive convection currents. As the mantle stirred, Pangaea and the surrounding oceanic crust began to break apart. A system of crustal plates developed. Molten material from beneath the crust nudged its

Earth has around 12 major tectonic plates; their boundaries are shown in red.

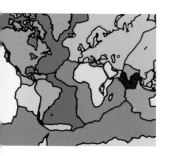

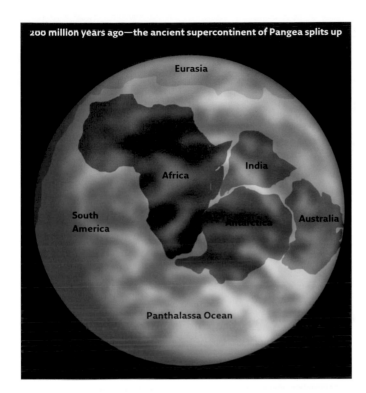

200 million years ago—the ancient supercontinent of Pangea splits up

Eurasia

Africa

India

South
America

Antarctica

Australia

Panthalassa Ocean

Around 200 million
years ago, the
ancient
supercontinent
of Pangaea began
to split up.

way into the gaps between the continents and cooled to form
new oceanic crust, and the land masses were forced to move
away from each other. The South American continent broke
away from west Africa, and they, along with India and Australia,
broke away from ancient Antarctica. Asia-Europe and North
America remained joined until around 100 million years ago,
when a fissure developed between Europe and North America,
creating an ever-enlarging Atlantic oceanic crust which has
widened at the rate of around one and a quarter inches per year.
Worldwide, the average rate of continental drift is 1in (2.5cm)
per year—about the same rate that your fingernails grow.
Although it appears to be an incredibly slow process in human
terms, the continuing drift of the continents can be directly
measured by using NASA's orbiting LAGEOS satellites.

Plate tectonics in action

Continents are ancient, crustal masses, largely made of lighter granitic rock augmented with layers of sedimentary and metamorphic rock; they average 25 miles thick. Oceanic crust is produced along huge volcanically active fissures on the ocean floor called mid-ocean ridges. Composed of dense basaltic material, the oceanic crust is far younger than continental crust, and is just six miles thick on average. Earth's continental and oceanic plates float upon a hot mantle stirred by giant convection currents which drive plate tectonics. There are around twelve major plates and a number of smaller ones; each plate is moving relative to its neighbors.

Subduction, volcanoes, and 'quakes

A process called subduction takes place where oceanic crust is being pushed towards a continental plate boundary—as the continent is crushed, its rocks are folded and uplifted into mountain ranges. The oceanic crust plunges beneath the continent, and as it is pushed into hotter parts of the mantle, the material remelts, rises, and intrudes into the lighter continental mass above. When this material breaks through to the surface, volcanoes are formed. Such a plate boundary can be found along the west coast of North and South America.

Subduction is not a smooth process: pressure gradually builds up at the plate boundary and is suddenly relieved with a jolt, producing earthquakes. These are common occurrences around the Pacific rim.

Mountain building

Where continents collide, the crust crumples and vast mountain ranges are thrust up. The most impressive example of such an occurrence is the Himalayan mountains, which began to be uplifted when India collided with Asia around 70 million years ago. The collision process continues to this day.

---

must know

There are 1,500 active volcanoes around the world. Around 150 strong earthquakes (above magnitude 6) take place each year.

Mount St Helens in Washington State— a volcano produced from the subduction of the Pacific Ocean plate beneath the North American continental plate. The volcano blew its top in 1980.

A chip off the third rock

Measuring a quarter of the Earth's diameter, the Moon is our only natural satellite. Whatever its origin, the Moon has been intimately linked with the Earth for billions of years—some even call it Earth's sister planet.

Now discredited, the fission theory of the Moon's origin saw the Moon as a small nodule which split off from a fast-spinning proto-Earth.

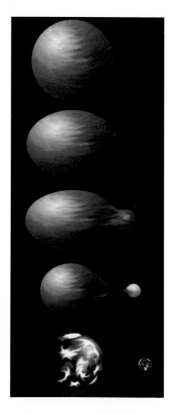

Whence the Moon?

The Moon's origin has been the subject of scientific speculation since the late 19th century, when George Darwin suggested that the Moon might have been a piece of the newly-formed Earth, which was flung off because of our planet's high rate of spin. Another theory deemed unsatisfactory is the proposal that the Moon was once an independent planet which was gravitationally captured by the Earth in the remote past.

It might seem logical to assume that the Moon was formed from the same cloud of primordial gas and dust which formed the Earth, but the Moon's composition is so dissimilar to that of the Earth that it is difficult to devise a scenario that sees it being formed in the vicinity of our planet at the same time as the Earth's formation.

The Big Whack theory

It may sound highly implausible, but the most widely accepted theory of the Moon's origin assumes a Mars-sized planet struck the young Earth a glancing blow, throwing out a massive plume of melted material. While the impactor was destroyed in the collision (its heavier core being absorbed into the body of the Earth), much of the material blasted

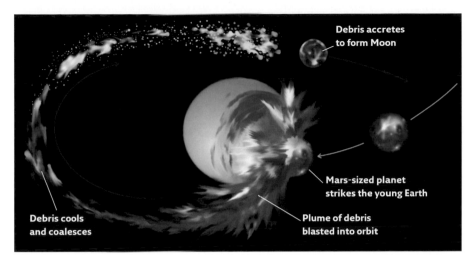

Debris accretes to form Moon

Mars-sized planet strikes the young Earth

Plume of debris blasted into orbit

Debris cools and coalesces

into orbit around the Earth coalesced and gravitationally pulled itself together into the Moon.

Bombarded Moon

Like the Earth, the Moon was on the receiving end of countless meteoroid, asteroid, and comet impacts. While the Earth has been a constantly dynamic world, its surface continually being eroded and deformed by a wide variety of geological processes, the Moon still clearly displays the scars of its bombardment. Carved into the solid lunar crust, impact craters dating back more than three billion years can be seen on the Moon's surface. The Moon was too small and its gravity too insubstantial for it to hold onto any gases that it may have generated by its once-active volcanoes. Having cooled faster than the Earth, volcanic activity on the Moon had dwindled by around three billion years ago. Incredibly, the dark lava plains formed during this distant era can be seen with the unaided eye.

Currently favored, the Big Whack theory of the Moon's origin involves an immense collision between the young Earth and a Mars-sized planet around 4.5 billion years ago.

Big moon

At 2,160 miles (3,476km) across, the Moon is larger than every planetary satellite in the Solar System, with the exception of Saturn's Titan and Jupiter's Io, Ganymede, and Callisto. The Moon is huge in comparison with the size of the planet around which it revolves—only Charon, Pluto's satellite, is proportionately larger.

Lunar synchronicity

Orbiting the Earth at an average distance of 239,000 miles (384,401km), the Moon is in what is known as a synchronous rotation, meaning that it revolves on its own axis in exactly the same time that it takes to complete one circuit around the Earth. As the Moon orbits the Earth every month, it keeps the same sphinx-like face turned toward us—as a consequence, an entire hemisphere of the Moon remained completely unknown until the Russian spaceprobe Luna 9 first photographed it in 1959.

Big and bright, the full Moon illuminates the night sky.

Stunning phases

Each month the Moon appears to change shape from crescent, to half Moon, full Moon, back to half Moon, and crescent again. This cycle of illumination, known as the Moon's phases, is caused by the changing amount of the Moon's surface we can see lit up by sunlight. Of course, apart from those rare occasions when the Moon enters the Earth's shadow and becomes eclipsed, the side of the Moon's surface facing the Sun is always fully illuminated, while the other hemisphere is in complete shadow. Since the Moon takes around four weeks to complete each orbit around the Earth, we observe a regular cycle in the amount of illumination exhibited by the Moon. The Moon's cycle of phases proceeds from New Moon, when the Moon lies between the Sun and the Earth, through its crescent phases, to half Moon (known as First Quarter), about a week after New. The Moon then becomes a gibbous phase and fills out to become Full after a further week; at Full Moon the Moon lies directly opposite to the Sun in the sky. Following Full Moon, the illuminated portion of the Moon narrows as the Moon becomes gibbous; half Moon (Last Quarter) takes place around a week after Full, and the Moon then assumes an ever narrowing crescent phase as it approaches New Moon once more.

One face, two sides

We now know that the Moon has two distinctly different sides. The familiar near-side is made up of bright cratered highlands and a largely interconnected set of large, relatively smooth dark plains. Known as the *maria* (Latin: seas), these dark patches represent large expanses of lava which flowed into huge asteroid impact craters several billion years ago. Of course, the Moon has long since cooled down—the lavas have solidified, and there's not a trace of volcanic activity on the Moon today. In contrast, the Moon's far-side has very few smooth lava-flooded areas—most of it is rough and pockmarked with craters.

Portrait of the Earth and Moon taken by the Galileo spaceprobe from a distance of 4 million miles (6 million km) in 1992.

Craters of a variety of shapes and sizes dot the Moon's rugged surface.

Pockmarked Moon

Most of the Moon's craters were formed by the impacts of asteroids, meteoroids, and comets in the first couple of billion years after its crust had solidified. Older craters appear more battered and eroded, and fresher craters are superimposed upon them. A number of prominent young impact craters are surrounded with magnificent bright rays of material thrown out by the impact explosion. Without an atmosphere, the Moon has no wind or rain, and erosion of features only takes place slowly by the impact of small micrometeorites which sandblast the solid crust over the aeons. To give some idea of the slow pace of this erosion, it is probable that the astronauts' footprints, left at six sites on the Moon's near side between 1969 and 1972, remain visible today just as crisply defined in the gray lunar soil as if they had been made yesterday.

Sculpted crust

A variety of fascinating geological features can be found carved into the solid lunar crust. Many of the Moon's mountain ranges rise to impressive heights above their surroundings—some

A magnificent rift valley slices through the lunar Alps.

peaks in the lunar Apennines, for example, reach heights of more than 16,500ft (5,000m). Ranges like the lunar Apennines, Alps, and Carpathian mountains were formed directly as a result of the impact that produced the large asteroid impact basin of Mare Imbrium (the Sea of Rains) around which they skirt.

Low, rounded hills known as domes can be found here and there, some of them topped by small craters. Domes are thought to be long-extinct lunar volcanoes, complete with summit vents. Indeed, lava flows and small valleys produced by flowing lava can be seen winding their way down the flanks of some domes.

Finding faults

Some parts of the Moon's crust have experienced a degree of tension, a pulling-apart movement which has caused the crust to crack along fault lines. A remarkable fault known as the "Straight Wall" forms a giant cliff face some 68 miles (110km) long, dividing the eastern part of Mare Nubium (Sea of Clouds). Fault valleys called graben-rilles have been produced where two close parallel faults have allowed the part in between them to sink down. Narrower rilles curve around the edges of some maria and within certain craters.

By far the largest of such features is the Alpine Valley, a giant crustal rift 80 miles (130km) long and 11 miles (18km) wide in places, which slices clean through the lunar Alps.

Faults and rilles are also common in and around some of the larger lunar craters. An impressive linear rille cuts cleanly across the floor of the crater Petavius, from its central mountains to its inner ramparts, and extensive networks of rilles can be seen criss-crossing the floors of the craters Gassendi, Posidonius, and Alphonsus, to name but a few examples.

want to know more?

Take it to the next level...
▶ **A perspective on the Earth and Moon in space** 12–13
▶ **Scale of the Earth-Moon system** 18
▶ **How our ancestors saw the Moon** 25

Other sources...
▶ **Explore your neighborhood and countryside—your landscape was molded by geological processes, and its rocks have a story to tell.**
▶ **Go look at Luna! Binoculars are cheap, and viewing the Moon is priceless.**
▶ **Telescopic views of the Moon will show you incredible detail.**

Weblinks...
▶ **An excellent resource for everyone wanting to get to know the Earth www.geology.com**

▶ **The Society for Popular Astronomy's Lunar Section helps you discover the Moon www.popastro.com**

4 Our cosmic backyard

Arrayed between its blazing solar hub and its distant frozen worlds halfway to the nearest star, the Solar System contains an astounding collection of individual worlds. Each world tells its own history—some give clear and direct accounts of themselves, while others speak to scientists in cryptic whispers. This chapter offers a glimpse into these worlds.

Other worlds than ours

Telescopes show the Sun, the Moon, and the major planets in considerable detail. A number of planets and their moons, asteroids, and comets have been visited by spaceprobes, providing stunning views of alien landscapes and wild atmospheres.

Terrestrial planets

Four large rocky worlds—Mercury, Venus, Earth, and Mars—dominate the inner Solar System. Venus, Earth, and Mars have substantial atmospheres, each strikingly different, but only the Earth has conditions suitable for liquid water to exist on its surface. These four worlds, known as the terrestrial planets, all display a variety of geological processes, including crustal movement and tectonics, and their surface features have been variously produced by impact cratering, volcanic activity, faulting and folding, erosion and sedimentation.

Minor planets

A wide zone of space beyond Mars is occupied by the main asteroid belt. Ceres, a cratered lump of rock 620 miles (1,000km) across, is the largest of these so-called minor planets, but most of them are considerably smaller.

Gas giants

Four immense planets—Jupiter, Saturn, Uranus, and Neptune—orbit the Sun beyond the main asteroid belt. Each of them is swathed in a deep layer of mainly hydrogen gas which is compressed into its liquid state deep below the cloud tops. Jupiter, the largest, is so big that a thousand Earths could comfortably fit within its volume. Saturn is surrounded by a beautiful ring system—a myriad of individually orbiting objects ranging from ice chunks the size of houses to tiny dust grains.

Uranus and Neptune, both discoveries of the telescopic era, are about four times the diameter of the Earth.

Ice worlds at the fringes

Pluto, an object smaller than the Moon but classed as a planet since its discovery in 1930, is one of a growing number of sizeable objects lying beyond the orbit of Neptune. In July 2005, an object much larger than Pluto and three times its distance—provisionally designated 2003 UB313—was heralded as the Solar System's tenth planet.

Deep frozen debris

Comets are thought to be made up of debris that has been left over from the Solar System's formation. Each is made of a combination of rock and ice, and they range in size from a few miles to several tens of miles.

Major objects in the Solar System, sized to scale.

Daytime star

By far the biggest and most important object in the Solar System, the Sun lies at the center of an amazing array of planets, asteroids, and comets. The Sun's light illuminates its family, and its energy drives planetary weather systems and sustains life on Earth.

watch out!

It is dangerous to view the Sun directly through binoculars or a telescope—a fraction of a second of magnified sunlight can cause permanent blindness.

Sunshine

Viewed from the Earth at a distance of 93 million miles (150 million km), our nearest star—the incandescent globe we call the Sun—is blindingly brilliant. It may look pretty small from the Earth's surface, but it measures an impressive 870,000 miles (1.4 million km) across; its volume is large enough to accommodate a million Earths. A process known as thermonuclear fusion takes place at the Sun's core, where hydrogen is converted to helium, and a little mass is lost in the form of energy. The Sun has been converting some four million tonnes of mass into energy each and every second for more than 4.5 billion years. There is no danger that the Sun will run out of fuel—it is so big that the process will continue vigorously for billions of years to come.

Sunspots can often be seen on the Sun's surface.

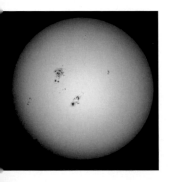

Solar activity

A powerful magnetic field generated within the Sun causes disturbances on its glowing outer surface, notably sunspots. While the solar surface has a temperature of 5,500°C, the interior of a sunspot is around 2,000°C cooler. As the Sun rotates on its axis every four weeks or so, the sunspots appear to slowly drift across the solar disk from one edge to the other.

Special equipment shows a zone of activity above the visible surface of the Sun, where large loops and jets of gas twist high above the solar surface. Some of these prominences hang there for days on end, while others shoot away from the Sun at a tremendous velocity.

Solar cycle

The Sun has an 11-year sunspot cycle, where the numbers of sunspots rises and falls on a regular (but not absolutely predictable) basis. When the Sun is at its least active, sunspots may be absent for several weeks at a time. At sunspot maximum there is a proliferation of spots, with occasional giant sunspots which are far larger than the Earth. The next solar maximum will take place around 2011.

A curving tongue of gas more than 620,000 miles (one million kilometers) long streams away from the Sun.

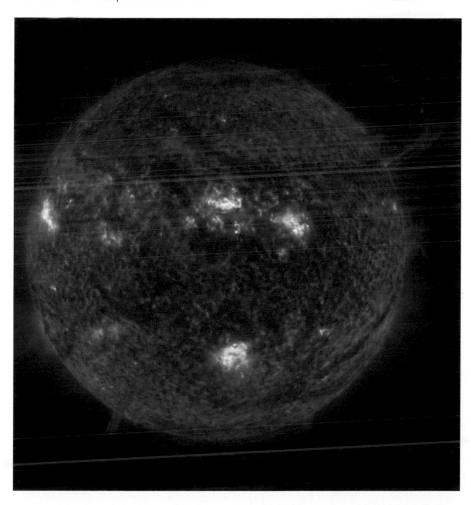

Mercury

Despite its relative nearness to the Earth, little Mercury, the innermost planet, remains one of the least-known objects in the Solar System. Only a third of its surface has been imaged by spaceprobes.

must know

Inside track
Mercury zips around the Sun at speeds of up to 130,000 miles (212,000km) per hour, completing each orbit in just 88 days.

Battered and crumpled

Measuring just 3,000 miles (4,880km) across, Mercury is about two-fifths the size of the Earth. Our only close-up view of this rocky, Sun-baked planet was provided by the Mariner 10 spaceprobe as it flew past the planet on three occasions in 1974 and 1975. The images showed a heavily cratered surface remarkably similar to some of the Moon's cratered highlands.

Consisting of ancient, highly eroded craters overlain with fresher brighter ones with rays, the Mercurian landscape bears witness to a long history of bombardment from asteroids—an era which probably lasted from the time the planet was formed to around three billion years ago. Large ridges, scarps, and fractures in Mercury's crust are thought to have been caused when Mercury's core cooled and shrank, buckling the overlying crust just like the wrinkled skin of an old apple. Interestingly, Mercury may have undergone more recent volcanic activity— small, smooth plains looking like lava flows were photographed here and there by the Mariner 10 spaceprobe.

Mercury's largest single feature is a giant asteroid impact scar called Caloris Planitia, 810 miles (1,300km) across. The immense shock waves produced by the impact cracked and deformed the crust, producing a feature which has the pattern of cracks in a car windshield holed by a bullet. After the Caloris impact, the crust slowly adjusted, and steep cliffs appeared along the fault lines; molten material bubbled from beneath the planet's crust, filling in some of the low-lying areas with lava flows.

Only the merest trace of an atmosphere surrounds Mercury—it is virtually a vacuum, so the planet experiences no weather. Without an appreciable atmosphere to distribute heat around the planet, there is a great temperature difference between the planet's day and night sides. Midday temperatures at the equator reach 350°C, while the night side plunges to around −180°C—the most extreme temperature range of any planet.

Crescent Mercury, imaged by the Mariner 10 spaceprobe.

Venus

Although Venus is a terrestrial planet almost as large as the Earth, its atmosphere, surface conditions, and geology could not be more different from those of our own planet.

The Venusian landscape, modeled on radar data returned by the Magellan spaceprobe.

Shades of Hades

Named after the Roman goddess of Love, Venus often appears as a beautiful bright star-like object in the morning or evening skies. Its substantial atmosphere hosts such a thick blanket of clouds that its surface can never be observed visually, and for centuries astronomers speculated about what

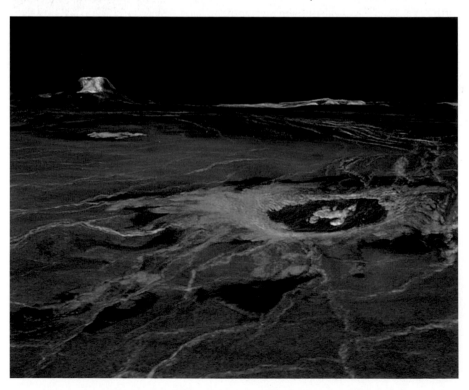

conditions might be like on its surface. Some thought that beneath its bright clouds, Venus might be a hot, humid world, its surface covered with lush tropical vegetation among which lurked a strange collection of Venusian life-forms. Others thought that Venus was covered with vast oceans fizzing with carbon dioxide gas—seas of foaming soda water.

Spaceprobes have shown that the surface of this once mysterious planet is incredibly inhospitable—despite the planet's proximity to the Earth, it seems certain that humans won't set foot upon the planet for a long time. Floating in an atmosphere of carbon dioxide, Venus' clouds are made of sulphuric acid. On the surface, the air pressure is 90 times that on Earth at sea level—equivalent to being 3,300ft (1,000m) beneath the sea. Trapping the Sun's energy, Venus' atmosphere maintains a runaway greenhouse effect, keeping surface temperatures a blistering 460°C—about 200°C hotter than a kitchen oven at its maximum setting. It's a scenario that closely matches traditional visions of hell.

Venus is covered by rolling plains, from which rise several large plateaux and two impressive continent-sized mountain ranges—Ishtar Terra and Aphrodite Terra. Unlike the Earth, Venus' crust is not split into drifting plates, and its sprawling mountains have been built up by billions of years of volcanism.

Venus' highest mountain, Maxwell Montes, towers 7 miles (12km) above the plains surrounding Ishtar Terra, and is more than 2 miles (3km) higher than Mount Everest. Aphrodite Terra boasts Venus' deepest valley, Diana Chasma, a giant canyon around 175 miles (280km) wide and 2½ miles (4km) deep in places.

Venus' thick swirling clouds, imaged by the spaceprobe Mariner 10.

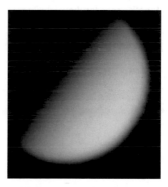

Venus' phases can clearly be viewed and imaged from the Earth.

Mars

More scientific speculation and flights of fictional fancy have centered around Mars than any other planet. Of all the worlds in the Solar System, Mars is the most Earth-like. It is probable that humans will tread upon its red soil within the next few decades.

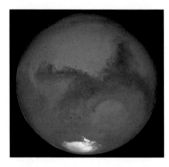

The red planet, imaged in August 2003 by the Hubble Space Telescope on the date of Mars' closest ever approach to the Earth.

Martian atmospherics

Mars is just over half the size of the Earth and has a surface area equivalent to the Earth's continents. Its day is only 37 minutes longer than our own, and its axis of rotation is tilted at about the same angle as that of the Earth, so Mars experiences a cycle of seasons during its 687 day long year. Mars' surface temperature averages an exceedingly chilly $-63°C$, but midsummer temperatures at the equator sometimes reach a tolerable 20°C. All the gases making up the Martian atmosphere can be found in the air we breathe on the Earth, but in different proportions. Carbon dioxide makes up 95 per cent of Mars' atmosphere, while oxygen accounts for only around one part in a thousand. Mars' atmosphere contains around a thousandth of the amount of water vapor present in the Earth's atmosphere, yet this is sufficient to cause wispy clouds to form at high altitudes. Bright clouds can often be found billowing in the lea of some of Mars' highest mountains, and fog patches sometimes form in low lying areas. On a small scale, dust-devils frequently whip up and zip across the Martian plains, while planet-wide dust storms smother the planet every few decades. They can be severe enough to completely mask Mars' surface features.

Mars' icy poles

Bright icecaps crown the Martian poles. Water ice forms the small core of both icecaps and is present all year round, while carbon dioxide ice accumulates around each pole during the wintertime, causing the caps to grow. When at its largest during the southern winter, the south polar cap is around 2,500 miles (4,000km) across and covers about 20 per cent of the entire surface area of Mars.

Recent observations by the Mars Express spaceprobe have revealed thousands of square miles of permafrost surrounding the south polar cap. This permafrost had evaded detection because of its dark color.

Bright clouds of water vapor, imaged over the Tharsis and Valles Marineris region.

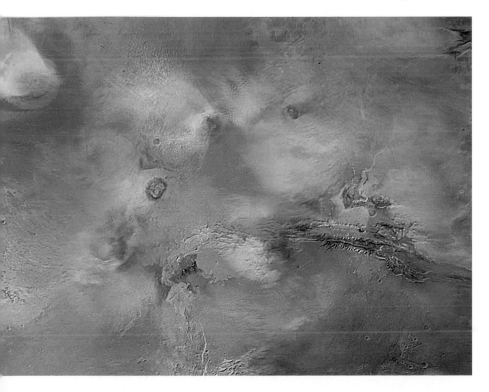

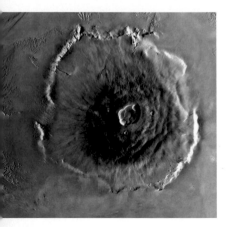

Mars' Olympus Mons is one of the Solar System's biggest volcanoes.

The Martian abode

Rocky deserts strewn with boulders and wind-blown dunes of red soil cover much of the Martian surface. Impact craters dot much of Mars' southern hemisphere, while the northern hemisphere is rather crater-free. This tells us that the northern hemisphere's surface has undergone considerable change since the era of major cratering in the Solar System several billions of years ago.

Mars' two distinctly different looking hemispheres can easily be noticed through a telescope. The planet's darker tracts mainly occupy its southern hemisphere in a near-continuous band encircling Mars. Among the most prominent of Mars' dusky patches are the V-shaped Syrtis Major and Solis Lacus, the "Eye of Mars." These dark mottled areas do not always conform to surface topography. There are many brighter features too, including the impact basins of Hellas and Argyre.

Valles Marineris, a giant rift valley system, extends almost a quarter of the way around Mars' equatorial region.

Just north of the Martian equator, four huge shield volcanoes rise from a giant bulge in Mars' crust known as Tharsis. Of these volcanoes (none of which have been active for many millions of years), Olympus Mons is by far the largest. Measuring 385 miles (620km) across and rising to a height of 17 miles (27km) above the surrounding terrain, Olympus Mons is three times higher than Mount Everest. A sizeable volcanic crater around 47 miles (75km) wide caps the mountain's summit.

A huge system of broad, deep valleys, known as Valles Marineris (named after the Mariner 9 Mars orbiter which imaged it in 1971), runs from the southeastern margin of the Tharsis plateau and extends a quarter of the way around Mars, just south of the equator. Valles Marineris is almost 3,100 miles (5,000km) long and is up to 5 miles (8km) deep. Even the smaller canyons which carve into the sides of the main valley would utterly dwarf the Grand Canyon in the United States. While the Grand Canyon has been formed by erosion from the Colorado River over millions of years, Valles Marineris is a rift valley created by the pulling apart of Mars' crust.

Spuds in orbit

Phobos and Deimos, both city-sized captured asteroids, are the two potato-shaped satellites of Mars. Orbiting at a height of just 1,600 miles (2,583km) above the planet, Phobos completes each circuit in 7h 39m—much faster than Mars rotates. Phobos is gradually approaching Mars and it will enter the Martian atmosphere in around 100 million years, impacting on its surface and carving out a giant impact crater.

Phobos, the innermost satellite of Mars, is an irregular shaped chunk of rock just 16 miles (26km) long.

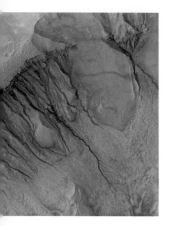

Liquid-cut gullies can clearly be seen in this image of a cliff on Mars.

Watery Mars

Today, Mars' low atmospheric pressure prevents water from existing as a liquid on the surface. When water ice is heated by the Sun it turns directly into vapor in the Martian atmosphere. However, there is no doubt that in the past Mars' atmospheric pressure has occasionally been high enough to allow water to flow across its surface, forming shallow seas and lakes which were fed by rivers.

There's plenty of evidence to back up this idea. Many of Mars' canyons and valleys appear to have been carved by flowing water, and there are also a number of streamlined features which appear to have been obstacles in the path of considerable water outflows. Mars was a great deal warmer and wetter several billion years ago; perhaps there have been cycles of warming and freezing.

Large quantities of ice have been discovered beneath the Martian surface, in areas far away from the poles, even along parts of the equator. Large accumulations of dust hide the ice well, but its presence has been confirmed by instruments onboard recent spaceprobes. Conditions in Mars before these deposits froze out were suitable for liquid water to exist in substantial quantities.

Martians

Primitive forms of Martian life may have developed in the remote past—indeed, it might exist today, hidden in Mars' soil. In 1976 the Viking landers tested for signs of life by scooping up soil samples and testing them chemically. Although it was claimed at the time that the tests were inconclusive, the inventor of one of the experiments maintains that the results were encouraging. In August 1996 NASA announced that signs of fossilized bacteria may have been found in a Martian meteorite—a rock that had been blasted off Mars' surface by an asteroid impact many millions of years ago.

In the search for life on Mars, the Viking landers dug and sampled small trenches in the red soil.

Asteroids

Hundreds of thousands of asteroids—chunks of rock ranging from the size of houses to the size of California—orbit the Sun. Most of them pose no threat to the Earth.

Crowded skies

Of those orbiting in the main asteroid belt between the Earth and Mars, asteroid Ceres is the largest, with a diameter of nearly 620 miles (1,000km), and there are 26 asteroids larger than 125 miles (200km). Surprisingly, if all the known asteroids were gathered together they would form an object just half the size of our own Moon.

Piloting the Millennium Falcon through the Hoth asteroid belt in the movie *The Empire Strikes Back*, Han Solo used all his skills to avoid colliding with the thousands of asteroids in his path. That's how many people imagine the asteroid belt to be, but things are different in reality. When real spacecraft are sent through the Sun's asteroid belt, mission controllers have no concerns that their billion dollar probes will accidentally crash into anything. Plotted on a graphic, the hundreds of thousands of known asteroids make the main asteroid belt appear incredibly crowded, but the asteroids are spread throughout such an enormous volume of space that collisions are few and far between. In fact, if you were standing on an asteroid, the nearest asteroid would likely be a distant starlike point of light.

Potential hazards

Throughout its 4.6 billion year history, the Earth has received countless meteoroid and asteroid impacts. It has been estimated that during the past half billion years alone the Earth has been hit by around 2,000 asteroids. Traces of these ancient impacts have largely been obliterated by surface erosion and sedimentation and deformed by the movement of the Earth's crust, but more than a hundred impact craters have been identified—including a 125 mile (200km) diameter crater buried beneath the Yucatan Peninsula, caused by a huge impact that is thought to have brought about the demise of the dinosaurs 65 million years ago.

There is a possibility that the Earth will be impacted again by a large asteroid, but nobody knows when this might occur. Such an asteroid will probably come from one of several classes of near-Earth asteroids, known as Potentially Hazardous Objects, more than 700 of which have been identified in a concerted effort to assess the risks. Many more of these objects remain to be discovered, but the general consensus is that the risks of a major impact in the near future are exceedingly small.

This artist's conception shows a potentially hazardous asteroid skimming past the Earth.

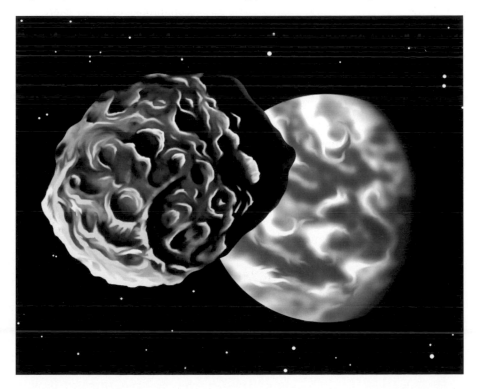

Jupiter

With a diameter of around a dozen times that of the Earth, Jupiter is so voluminous that all the other planets and their satellites, plus all the known asteroids and comets in the Solar System, could easily fit inside it, with lots of room to spare.

must know

Great Red Spot
The nearest thing to a permanent feature on Jupiter is its Great Red Spot—a particularly long-lived oval-shaped anticyclone that may have existed for at least 350 years. Larger than the Earth, the Great Red Spot has no physical existence other than the swirling winds which feed and sustain it.

Giant planet

Jupiter is a gas giant composed largely of hydrogen and helium gas, plus a few other elements. Its composition is thought to be similar to the original nebula from which the Solar System was born around five billion years ago. With an incredibly fast rate of spin of less than ten hours, Jupiter has a pronounced equatorial bulge caused by centrifugal forces, and its cloud features are stretched out into a series of bright and dark bands parallel to the equator.

Spots and ovals frequently bubble within the Jovian atmosphere, making a fascinating, constantly changing scene. It was once thought that there might be a solid surface deep beneath the planet's atmosphere—a landscape of gigantic proportions with towering volcanoes—but it is now known that Jupiter does not have a solid surface, and no spaceprobe could ever make a landing there. Beneath Jupiter's cloud tops, the atmospheric pressure increases so much that gases are compressed into liquid form. Its core may be a giant ball of liquid rock bigger than the Earth.

Jupiter's atmosphere displays banding with fantastic swirls, eddies, and spots. The Great Red Spot looms large in the planet's south tropical zone.

Failed star?

Many planets of Jupiter's mass and larger have been discovered orbiting other nearby stars. It is often said that Jupiter is a failed star—had it been more massive, the high internal pressures and temperatures would have triggered nuclear reactions, igniting the hydrogen and causing it to shine like a star.

Giant Jupiter is an unforgettable telescopic sight.

The Galilean moons

Known as the Galilean moons after their discoverer, Galileo Galilei, Jupiter's four largest satellites are sizeable worlds in their own right. Ganymede and Callisto are both bigger than Mercury; Io is larger than our Moon, while Europa is slightly smaller.

If a good illustration of the workings of gravity and Kepler's orbital laws is required, all that's necessary is to watch the movements of the Galilean moons over the course of a few days.

Io

Jupiter's innermost satellite has a patchwork of yellow, orange, red, brown, and black markings that gives it a pizza-like appearance. There are few signs of impact cratering, just a jigsaw of smooth plains, crustal cracks, and strange mountains topped with irregular volcanic craters. Pulled around by the gravity of Jupiter and the other Galilean moons, Io's insides are heated by tidal friction and its solid crust deforms, fractures, and experiences volcanic activity. Plumes of material are frequently spewed high above its surface from a number of active volcanoes.

Europa

Europa's smooth icy crust is crisscrossed by hundreds of narrow dark lines, thought to be rifts which have been filled with warm icy slush from within its interior. A deep ocean of water may exist beneath its crust, and it has been speculated that this is of the most likely places in the Solar System for primitive life to have developed.

Ganymede

Ganymede, the largest satellite in the Solar System, is an icy world, like Europa, but its surface is older, pockmarked with impact craters, and shows signs of widespread crustal movement which has caused fractures, ridges, and folds in the crust. Darker regions on Ganymede contain more impact craters, suggesting that this is the satellite's oldest terrain.

Callisto

Craters and other impact features are just about the only things visible on Callisto, the outermost Galilean moon. It has the most heavily cratered surface of any place in the Solar System, its crust dating back at least four billion years, almost to the time of the Solar System's formation.

Io, Europa, Ganymede, and Callisto, shown to scale.

Saturn

Surrounded by its beautiful broad rings, gas giant Saturn is one of the most stunning sights in the Solar System, and its big satellite Titan is one of the most intriguing worlds yet explored.

Planetary lightweight

Saturn is the Solar System's second-largest planet, a rapidly spinning ball of mainly hydrogen and helium gas. It spins once on its axis in just over ten hours, and because it is somewhat less dense and has a lower gravity than Jupiter, it appears more flattened at the poles—its equatorial bulge, thrown out by centrifugal forces, produces a planet whose polar diameter is only 90 per cent its equatorial diameter. Saturn's overall density is the lowest of any planet. In fact, it is less dense than water, so if you could find an ocean big enough, Saturn would float in it (though its rings would melt and sink).

Saturn and its glorious ring system, imaged by the Cassini spaceprobe in 2005.

Atmospheric activity

Although Saturn's yellow colored atmosphere is banded with several dusky cloud belts, there is much less obvious activity than in Jupiter's atmosphere. This isn't surprising, because Saturn is more than 400 million miles (650 million km) farther from the Sun than Jupiter, and receives considerably less heat to drive its atmospheric processes.

must know

Occasionally, Saturn's belts develop short-lived irregularities and dark spots, and once every few decades a major atmospheric disturbance breaks into the atmosphere, producing big, bright spots which last for several months.

Rings

With their small, low magnification telescopes, early observers were puzzled by Saturn's appearance—it seemed to have two handle-shaped appendages which varied in size over a few years. In 1659, Christiaan Huygens correctly perceived the true cause of Saturn's odd appearance, writing that the planet was encircled by "a flat ring nowhere touching the planet." The changing tilt of Saturn causes the rings to appear to open up, narrow to a line and then open up again—a cycle that takes place during Saturn's 29-year orbit around the Sun.

Seen from the Earth, Saturn's main rings look solid—they are opaque and cast dark shadows onto the planet's globe. But a solid ring could not exist around Saturn (nor around any other planet), because gravitational forces would immediately tear apart such an object. Instead, the rings of Saturn are composed of countless billions of individual moonlets, ranging in size from dust grains to house-sized chunks. Here and there the rings bend and ripple as their particles resonate with the gravity of several small satellites in and about the rings. A narrow outer ring even appears braided and twisted around on itself.

Saturn's many moons

A fascinating collection of no fewer than 34 currently known
satellites orbit Saturn, including a giant satellite with its own
atmosphere, several medium-sized satellites with amazingly
varied terrain, and a number of sizeable but irregular-shaped
chunks of rock and ice. A number of small satellites known
as "shepherds" orbit in and among the rings, their gravity
keeping the rings in trim.

Titan

Titan, the largest of Saturn's moons, measures more than 3,200
miles (5,100km) across and is a giant planetary satellite, second
only to Jupiter's Ganymede in terms of size. Titan is the only
satellite known to have its own substantial atmosphere—a
dense yellow smog of mainly nitrogen and methane gas, which

**Surface detail on
smoggy Titan can be
seen in this
enhanced image,
taken by the Cassini
probe in 2005.**

completely hides Titan's surface from view. NASA's Cassini spaceprobe has revealed tantalizing glimpses of Titan's cold, icy surface by looking at it in different wavelengths of light and mapping it with radar. In 2005 the Huygens probe touched down on Titan's surface and imaged a rolling orange landscape covered with rounded ice chunks. Networks of drainage channels, cut by running liquid, run from bright highlands to smoother dark regions. Icy volcanoes have been discovered— large mounds built up of eruptions of slushy material oozing from beneath the surface. A few small young impact craters have been imaged, but Titan is such a dynamic world that new impact features are quickly eroded.

Battered icy moons

Saturn's medium-sized satellites—Mimas, Enceladus, Tethys, Dione, Rhea, and Iapetus—are all spherical worlds ranging from 250 to 930 miles (400 to 1,500km) across. Composed mainly of water ice, all display extensive impact cratering.

Mimas is densely cratered and dominated by a large impact crater called Herschel, more than one-third of Mimas' diameter —the impact very nearly destroyed the satellite.

Enceladus has a varied surface with many distinct kinds of terrain. Enceladus may be host to icy volcanism, with warm icy slush through cracks in the crust, which spread out to cover large tracts of the surface.

Tethys is covered by impact craters and fractures. A vast fissure called Ithaca Chasma, 1,200 miles (2,000km) long, over 40 miles (60km) wide and 4 miles (6km) deep in places, stretches from pole to pole, and a huge impact crater called Odysseus dominates one hemisphere of Tethys.

Dione and Rhea are similar looking moons. Both have been heavily impacted over a long period of time. Bright wispy streaks cover parts of both moons, and both are gravitationally locked to Saturn, so that the same side always faces the planet.

Mimas' remarkable appearance has earned it the nickname of the "Death Star" moon.

Uranus

Gas giant Uranus orbits at an average distance of 1.8 billion miles (2.9 billion km) from the Sun—twice the distance of Saturn. It takes 84 years to complete each orbit.

Uranus presented a bland, turquoise face to the spaceprobe Voyager 2 in 1986.

Green giant

Around four times the Earth's diameter, Uranus has a thick hydrogen and helium atmosphere which overlies a substantial core of ice and rock. Uranus' axis is almost in line with the plane of its orbit around the Sun, quite unlike other planets. It is possible that a giant impact in the remote past knocked the planet onto its side.

Uranus' turquoise hue is attributable to methane in its atmosphere. Since the visit of Voyager 2 in 1986, Uranus has gradually turned its equator sunward; being more evenly heated, the intensity of the planet's cloud bands, and activity within them, appears to be increasing.

Remarkable satellites

Only a few of Uranus' 27 known satellites were imaged up close by Voyager 2, the only spaceprobe to have visited the planet. Titania, Uranus' largest satellite, is more than 930 miles (1,500km) across. Its icy crust has experienced impact cratering and a lot of tension and faulting, producing large cliffs and fault valleys which wind across the satellite's surface.

Oberon has a mottled brown surface composed of ice and rock. Patches of darker material can be seen in many of the satellite's impact craters—this may consist of organic slush extruded from beneath the crust. A spectacular feature is Mommur Chasma, a large fault valley stretching halfway around the planet, produced when the crust was pulled apart.

Umbriel, the darkest of Uranus' major satellites, has a highly cratered surface; the largest, Wokolo, measures more than 125 miles (200km) across.

Ariel, a slightly less than round icy body, displays a complex history of impact cratering, icy volcanism, crustal melting, and large-scale ice movement, along with extensive faulting and valley formation.

Miranda is a strange potato-shaped ice world, displaying a varied surface that speaks of a complex history involving crustal movement, icy volcanism, and countless impacts. Some regions are crossed by faults, valleys, and giant angular cliffs, the steepest of which rises 11 miles (18km) above its surroundings. Circular concentric grooves and ridges are among the more unusual forms of terrain on Miranda.

Elusive rings

Uranus is encircled by a faint ring system unlike that of Saturn. It consists of nine thin, narrow, and widely separated components, the outermost of which is made up largely of ice chunks up to 3ft (1m) across and shepherded into shape by Uranus' small moons Ophelia and Cordelia.

Uranus' main satellites—Miranda, Ariel, Umbriel, Titania, Oberon— shown to scale.

Neptune

The outermost gas giant orbits the Sun at an average distance of
2.8 billion miles (4.5 billion km)—30 times the distance of the Earth
from the Sun. It takes 165 years to complete each orbit. On May 29th,
2011 Neptune will have made exactly one orbit since its discovery.

Wide blue yonder

Neptune is slightly smaller than Uranus, but the larger size of its
core gives it a greater mass. Neptune's 1,900 miles (3,000km)
deep atmosphere, composed mainly of hydrogen and helium, has
a high proportion of methane, giving the planet a distinctly blue
coloration. Prominent atmospheric bands girdle the planet.
Images taken by the Hubble Space Telescope show that there is
always a flurry of activity going on in Neptune's atmosphere,
including dark and bright spots, plumes, and festoons. Among the
atmospheric phenomena imaged during Voyager 2's brief flyby of
Neptune in 1989 were several large dark spots, including one
9,300 miles (15,000km) across, nicknamed the "Great Dark Spot."

Active Triton

Triton, one of the Solar System's biggest satellites, is the only
major satellite to orbit clockwise (viewed from above the planet's
north pole). Some areas appear smooth and featureless, while
other parts are wrinkled. Its surface is being renewed from
within; dark patches appear to be sites of local melting. Ice
volcanoes shoot geysers of nitrogen high above Triton, forming
dark streaks downwind. A cap of frozen nitrogen and methane
covers Triton's southern hemisphere.

Neptune's active atmosphere shows cloud belts, dark spots, and bright high altitude cirrus clouds in this image taken by Voyager 2 in 1989.

Tiny Nereid has the most eccentric orbit of any moon in the
Solar System. It approaches Neptune as close as 870,000 miles
(1,400,000km) and recedes as far as 600,000 miles
(9,600,000km) in a little under a year.

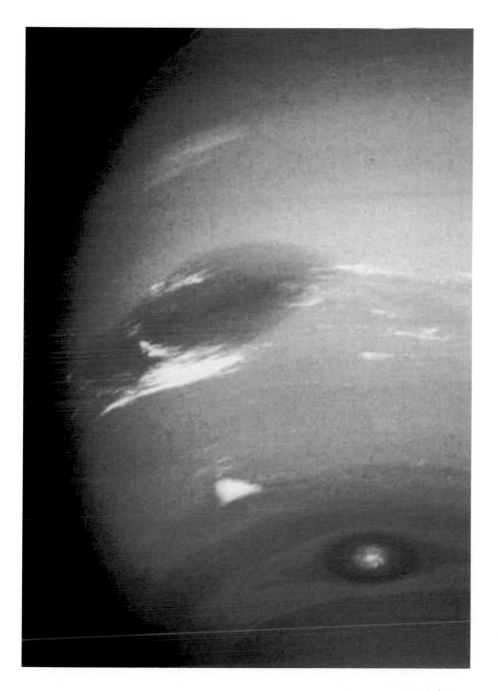

Worlds beyond Neptune

Pluto, the smallest and farthest of the Sun's nine "official" planets, was discovered photographically in 1930. In recent years, a considerable number of sizeable objects beyond the orbit of Neptune have been discovered.

Pluto, distant double planet

Measuring 1,400 miles (2,280km) across, Pluto is smaller than seven planetary satellites, including our own Moon. Little is known about this distant icy world. It remains the only major planet that has not been imaged at close range by a spaceprobe. Pluto takes 249 years to complete each orbit around the Sun. Its orbital path is more elliptical than any other planet, and the plane of its orbit is angled steeply to the general plane in which all the other major planets orbit. At its furthest, Pluto is more than 50 times farther from the Sun than the Earth.

Pluto's large satellite, Charon, is more than half the size of Pluto. In effect, the Pluto-Charon system is a double planet, since the pair orbit their common center of gravity in space. Pluto is thought to be a fairly dense mix of rock and ices, while Charon is more lightweight, composed chiefly of ices. Such a difference in their composition suggests that Pluto and Charon were formed independently of one another. Perhaps Charon was the result of a planetary collision, formed in a similar way to the "big whack" scenario of our own Moon's origin.

Kuiper Belt objects

Astronomers have identified a zone of space beyond Neptune which is populated by hundreds of minor planets—a second asteroid belt around the Sun, known as the Kuiper Belt. More than 800 Kuiper Belt objects have been discovered, ranging in size from a few tens to a few thousand miles in

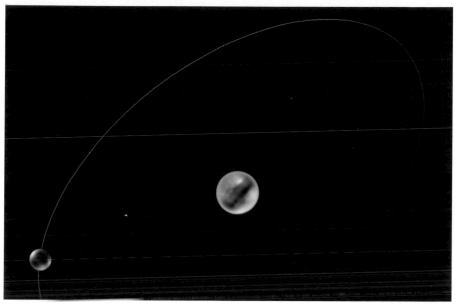

Pluto and its moon Charon, shown to scale.

diameter. If Pluto were discovered today, it would probably be classed as a Kuiper Belt object, rather than as a planet.

Planet X

The discovery of a Kuiper Belt object far larger than Pluto, temporarily named 2003 UB313, was announced in July 2005. Orbiting approximately three times farther from the Sun than Pluto in an orbit inclined by 45° to the plane of the Solar System, 2003 UB313 is spherical and may be twice as large as Pluto. The new object—trumpeted by NASA as the tenth planet—appears to have a surface covered with methane ice, like Pluto.

This Planet X is the largest object found in orbit around the sun since the discovery of Neptune and its moon Triton in 1846.

Comets

Comets are balls of ice, dust, and rock which formed from the material of the original solar nebula, far out in the deep freeze of the outer Solar System. Studies of comets and their composition tell us a great deal about the birth of the Solar System.

Comet Wild 2 has a highly pockmarked nucleus of dirty ice measuring 3 miles (5km) across. This image, a composite which includes the jets of dust and gas surrounding the nucleus, was taken during the January 2004 flyby of the Stardust probe.

The Oort Cloud

Far beyond the planets, in realms so distant that the Sun appears as a bright star-like point, there is a zone of space thought to contain trillions of comets. Known as the Oort Cloud (after Jan Oort who first suggested its existence in the 1950s), this zone stretches halfway to the nearest stars and marks the outer fringes of the Solar System. Individual comets within the Oort Cloud have never been directly observed, nor do we currently have much prospect of detecting them at such enormous distances. Although the precise size and shape of the Oort Cloud remains unknown, its inner region is speculated to be a flattened disk, broadly in line with the plane of the Solar

Brilliant Comet Hale-Bopp.

System; extending outward from around 50,000 AU, it broadens into a spherical outer zone, forming a vast shell extending halfway to the nearest stars.

Comets within the Oort Cloud are separated by many tens of millions of miles; the region is far more sparsely populated than the main asteroid belt between Mars and Jupiter. In the Oort Cloud's outer realms, the Sun's gravity has such a weak grip that the orbits of comets there can be perturbed by the gravity of nearby stars. As a result, they can be nudged into a different orbit within the Oort Cloud, flung off into interstellar space or enticed toward the inner Solar System, where we finally get to view them. Every half a billion years or so, the Sun, carrying the Solar System and the Oort Cloud along with it, passes through large interstellar gas clouds in the galactic arms. The gravitational attraction of these giant clouds distorts the Oort Cloud's weakly-bound outer shell, causing a significant disruption in the orbits of the outer Oort comets. As a result, more comets head into the inner Solar System, increasing the likelihood of collisions with planets. Cometary impacts have been cited as one possible cause of the periodic mass extinctions in the Earth's history.

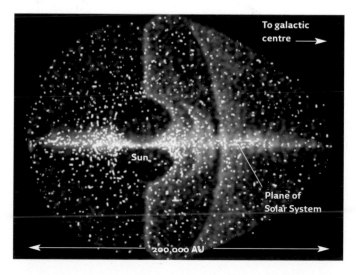

A cutaway section through the vast Oort Cloud of comets. The Solar System lies at the center—the orbit of the outermost planet Pluto is too small to depict at this scale.

To galactic centre →

Sun

Plane of Solar System

← 200,000 AU →

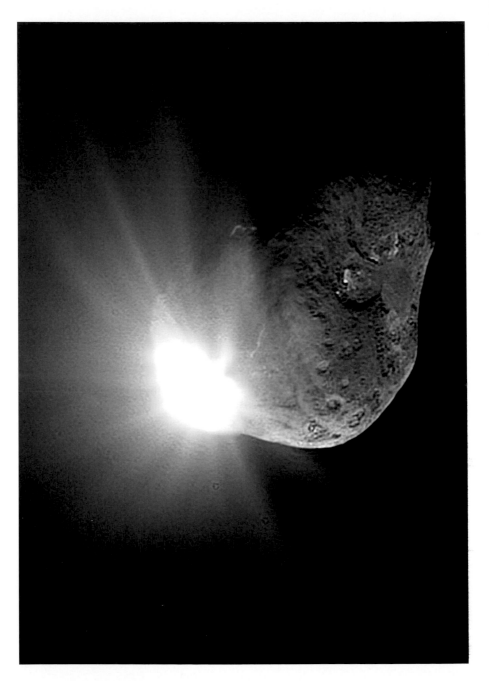

Familiars and fly-by-nights

Newly-discovered comets sweep unexpectedly into the inner Solar System, fresh from the Oort Cloud. Their orbits are so large that it is impossible to keep track of them as they fade and disappear into the depths of space, never to be seen again. However, there are 169 currently known "periodic comets" whose relatively small orbits around the Sun are known—their orbital periods range from 4 years (Comet Encke) to 400 years (Comet Ikeya-Zhang). Bright periodic comets form an impressive link with both the history of the Solar System and the history of humanity.

Heads and tails

Comets spend most of their lives far from the Sun as deep-frozen chunks of ice mixed with dust and rock. It's only when they approach the Sun, and their surfaces begin to warm up, that the action begins. Frozen material on the surface begins to sublimate— that is, it turns from a solid icy state into gas. As the gases drift away from the nucleus, bits of dust and rock once lodged in the ice are freed. The icy nucleus is soon surrounded by a coma—a reflective cocoon of dust and gas which hides the nucleus from view. As the comet gets closer to the Sun, activity becomes more vigorous, as jets of gas and dust shoot away from more volatile spots on the nucleus. The solar wind pushes the gas directly away from the Sun forming a straight gas tail, while solid debris forms a curving dust tail in the comet's wake. Each comet puts on a different display—some remain as rather faint fuzzballs, and are visible through binoculars only, while others sprout magnificent long bright tails, easily seen with the unaided eye.

In 2005 the spaceprobe Deep Impact lobbed an impacting probe into the nucleus of Comet Tempel 1—the impact produced a huge plume of bright material and gouged out a crater as big as a football field.

Hubble's planetary perspectives

Astonishing images of the varied worlds in our Solar System have been returned by the Hubble Space Telescope, whose large, ultrasensitive eye has peered into space from its orbit around the Earth since 1990.

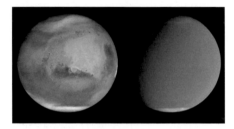

Between the dates that these images of Mars were taken—June and September 2001—Mars' surface had been completely obscured by a planetwide dust storm.

In 1994 dozens of chunks of the fragmented nucleus of Comet Shoemaker-Levy 9 slammed one by one into Jupiter's atmosphere, producing large dark markings. Jupiter's Great Red Spot is visible at upper right.

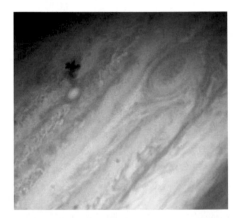

Montage showing the Hubble Space Telescope's views of the Solar System's four gas giants, sized to scale—Jupiter, Saturn, Uranus, and Neptune.

The fragmented nucleus of Comet Shoemaker-Levy 9, appearing like a string of jewels, imaged several months before its historic encounter with Jupiter.

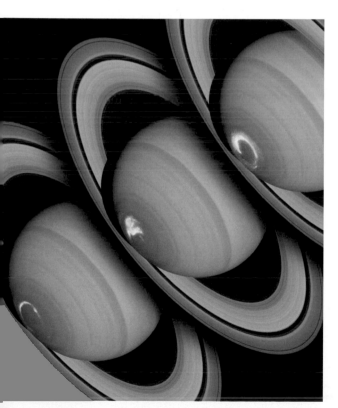

These three images of beautiful ringed Saturn, taken in January 2004, show the development of a bright auroral band above the planet's dusky south polar region.

want to know more?

Take it to the next level...
▶ Solar System in a nutshell 14
▶ Solar System in perspective 19
▶ Planets and pseudoscience 26-7
▶ Ancient notions of the Solar System's workings 36
▶ Understanding the Solar System 40-57

Other sources...
▶ You can view five planets with the unaided eye—buy a current astronomy magazine to find out where they are.
▶ A small telescope will show you the phases of Venus, the orange disc of Mars, the disc and moons of Jupiter, and the rings of Saturn.
▶ *Need to Know? Stargazing* is an excellent introduction to observing.

Weblinks...
▶ www.popastro.com Takes you to the Planetary Section of the Society for Popular Astronomy.
▶ www.nineplanets.org A super multimedia tour of the Solar System.

5 The galactic neighborhood

Our Solar System resides in a quiet suburb of the Milky Way galaxy, nestled within one of its spiral arms. A representative selection of stars lies within the Sun's immediate stellar neighborhood, from dim brown dwarfs to brilliant supergiant stars. Further afield, the Milky Way is full of star clusters, nebulae, and exotic objects such as pulsars and black holes.

Star colors

A star's color gives us an idea of its surface temperature. Our own yellow Sun has a surface temperature of around 6,000°C. Redder stars are cooler, while bluer stars are hotter.

Giants compared

A good color-temperature comparison is made by the two brightest stars of Orion, a constellation equally viewable from both the northern and southern hemisphere. Betelgeuse (Alpha Orionis) is distinctly red; the surface of this supergiant, a star 1,000 times the size of the Sun and 600 light years away, is around 2,000°C, about one third the Sun's temperature. Rigel (Beta Orionis) is a blue supergiant some 775 light years distant, and has a surface temperature of around 11,000°C.

Spectrum inspection

A respected scientist once maintained that the composition of the stars would forever remain a mystery to science. A few

Main sequence league table

Star type	Color	Surface temperature	Ave. mass (Sun=1)	Ave. diameter (Sun=1)	Ave. brightness (Sun=1)	Main sequence example
O	Blue	25,000°C +	60	15	1,400,000	Zeta Puppis
B	Blue	11,000°–25,000°C	18	7	20,000	Spica
A	Blue	7,500°–11,000°C	3.2	2.5	80	Sirius
F	Blue to White	6,000°–7,500°C	1.7	1.3	6	Procyon
G	White to Yellow	5,000°–6,000°C	1.1	1.1	1.2	The Sun
K	Orange to Red	3,500°–5,000°C	0.8	0.9	0.4	Pollux
M	Red	below 3,500°C	0.3	0.4	0.04	Proxima Centauri

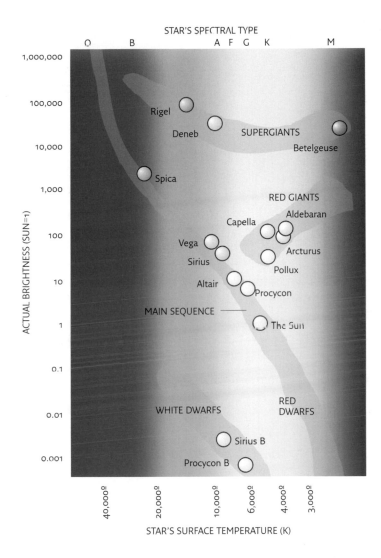

STAR'S SPECTRAL TYPE

O B A F G K M

1,000,000

100,000 — Rigel

Deneb — SUPERGIANTS

10,000 — Betelgeuse

Spica
1,000

RED GIANTS

Aldebaran
Capella
100 — Vega
Sirius — Arcturus
Pollux
10 — Altair
Procycon

MAIN SEQUENCE ———

1 — The Sun

0.1

WHITE DWARFS — RED DWARFS
0.01

Sirius B
0.001 — Procycon B

ACTUAL BRIGHTNESS (SUN=1)

40,000° 20,000° 10,000° 6,000° 4,000° 3,000°

STAR'S SURFACE TEMPERATURE (K)

In the Hertzsprung-Russell diagram, a plot of stars' spectral type from
O to M (hence their surface temperature) and their actual luminosity
produces a band called the "main sequence" from dim red dwarfs at
lower right, to huge ultrabright hot blue stars at upper left. All main
sequence stars are burning hydrogen to helium in nuclear fusion
reactions at their cores, and most stars spend about 90 per cent of
their life on the main sequence.

years later he was proved wrong when the spectroscope—a special piece of equipment used to analyze the makeup of light by splitting it into a spectrum—clearly revealed chemical elements in the stars. A multitude of thin dark lines in the spectrum of a star, known as absorption lines, are like the fingerprints of the elements.

More than 10,000 absorption lines have been identified within the spectrum of our nearest star, the Sun.

Spectroscopy allows stars to be studied in remarkable detail, even enabling the velocity of individual stars to be determined by means of a phenomenon known as the Doppler effect—a displacement of absorption lines toward the red end of the spectrum (redshift) shows that a star is moving away from the observer, while a displacement towards the blue end of the spectrum (blue shift) shows that a star is approaching the observer.

Spectral types

As we have seen, stars of different color have different temperatures. Stars of different temperatures have characteristic types of spectrum. These spectral types are identified by seven main groups—O, B, A, F, G, K, and M, from hot O-type blue stars, through G-type stars like the Sun to cool M-type red stars. An easy way to remember these types is by the oft-repeated mnemonic "Oh, be a fine girl/guy, kiss me!"

Star birth

Vast clouds of interstellar gas and dust within the galaxy's spiral arms—so dense and deep to be virtually opaque to visible light—

can be seen silhouetted against parts of the Milky Way. Dark nebulae may look pretty dull and inactive, but parts of them are gradually attaining just the right conditions to become the birthplace of stars. Over time, denser parts of the clouds may collapse under their own gravity; the process may also be triggered by a nearby supernova shockwave compressing parts of the cloud.

Within each contracting zone of interstellar dust and gas, a number of individual stellar birthplaces may appear. It may take around ten million years from the first stages of collapse to the appearance of an embryonic star, a region of unstoppable gravitational collapse called a protostar. Dust and gas attracted by the protostar's gravity produces heat, and pressure as its core rises. Temperatures at the core eventually become high enough to trigger thermonuclear fusion, and a true star is ignited. Dust and gas near the young star is blown away by a strong stellar wind, but any planets and larger chunks of debris that may have formed around the protostar will bathe in their new sun's light and energy.

Star development

All stars are not born equal. The amount of time a star spends on the Main Sequence of the Hertzsprung-Russell diagram depends on its mass—the larger the star's mass, the less time it will enjoy on the Main Sequence. A star 50 times the mass of the Sun will spend just a million years on the Main Sequence before it uses up all its hydrogen fuel. A star like the Sun will spend around ten billion years on the Main Sequence, while a star with only half the Sun's mass will spend 60 billion years on the Main Sequence.

Main Sequence stars show a relation between their mass and luminosity—more massive stars are brighter, since the fusion reactions at their cores take place at a faster pace than in less massive stars. So, despite starting out with more hydrogen fuel, more massive stars use it up more quickly and have shorter lives.

Turning off Main Street

As a star eventually uses up the hydrogen fuel at its core, its energy production falls and the core's temperature and pressure decreases. In response, the core contracts slightly, producing a sharp rise in temperature which in turn ignites the hydrogen shell surrounding the core—a zone that was previously too cool to undergo fusion reactions. This point marks the star's exit from the Main Sequence. As the star expands, its increased surface area makes it appear brighter, while its surface becomes cooler and redder. Main sequence stars which are more than 80 per cent the Sun's mass will become red giants, while much more massive O and B-type Main Sequence stars will turn into red supergiants.

Bright examples of red giants and supergiants in our cosmic neighborhood can be picked out with the unaided eye —their color tinge is obvious to many keen-sighted observers. Star colors tend to be more easily visible through binoculars, too. One of the brightest red supergiants, Betelgeuse, can be found in the constellation of Orion. Its diameter is about 1,000 times that of the Sun.

Puffing into oblivion

As the core contracts yet further, temperatures rise high enough to begin burning the remaining helium within the star's core, turning it into carbon ashes. As the core periodically contracts, the star's outer atmosphere is puffed into space as rings or shells of material known as a planetary nebula (because some examples of these objects resemble planets when viewed through a telescope eyepiece). At their center lies the highly compressed million degree stellar remnant—an Earth-sized object called a white dwarf, so dense that a thimble full of its stuff would weigh a tonne. Hydrogen gas within planetary nebulae glows as it is ionized by ultraviolet radiation emitted by the white dwarf. But planetary nebulae are short-lived—they are only visible for less than 100,000 years, as they expand and fade. Around a thousand planetary nebulae populate our part of the galaxy.

Although they shine for only a brief moment on the cosmic stage, planetary nebulae are among the most beautiful objects in the Universe. They have all been formed in the same way, yet these puffs of gas from dying stars vary enormously in appearance. Easily the largest and brightest of these, the Dumbbell Nebula in the constellation of Vulpecula, is visible through binoculars; it has two brightly glowing lobes, giving it the appearance of a luminous apple core. Another lovely bright planetary nebula, the Ring Nebula in Lyra, makes a beautiful sight through a small telescope—a luminous doughnut whose central white dwarf is just visible through a large telescope.

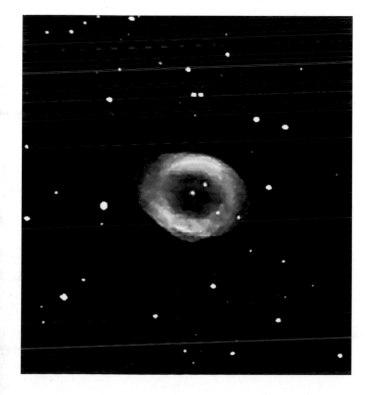

Glowing like a cosmic doughnut, the Ring Nebula in Lyra.

Stellar fates
The destiny of
individual stars
depends on their
mass and their
environment.
Some stars fade
into obscurity,
while others tear
themselves apart
in a catastrophic
supernova
explosion.

Supernovae

Really massive stars refuse to fade into oblivion. At the end of
their lives, and with their nuclear fuel exhausted, red
supergiants become highly unstable and their cores eventually
collapse. Within a fraction of a second, the core—an object the
size of the Sun—crumples in on itself by its own gravity. A shock
wave rebounds through the star's outer layers and a supernova
explosion ensues—a catastrophic event that produces so much
light that it can briefly outshine all the stars in its home galaxy.
These events are known as Type II supernovae. At the center,
the core's collapse may stop when it becomes a neutron star—
an object with about the same mass as the Sun but packed into
an ultradense object just six miles across. When first formed,
neutron stars spin at incredibly fast rates, and they emit radio
waves along their rotation axis—these are known as pulsars
because radio telescopes can detect regular pulses of radio
emission from them.

In 1054 AD, a brilliant new star suddenly appeared in the
constellation of Taurus. It was bright enough to be visible in

**Depending on the
mass of the star as it
goes supernova, its
core may crumple
into an ultradense
pulsar, or it may be
crushed out of
existence
altogether, forming a
black hole whose
gravity is so
powerful that
nothing—not even
light—can escape.**

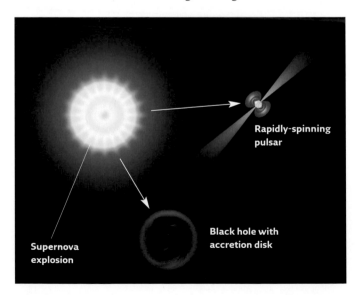

Rapidly-spinning
pulsar

Supernova
explosion

Black hole with
accretion disk

daylight for three weeks and remained a naked eye object for almost two years as it gradually dimmed. The remnant of this catastrophic supernova explosion—visible as a faint puff of gas through the telescope eyepiece—is called the Crab Nebula. Radio waves and visible light from the superdense, rapidly spinning pulsar at the center of the Crab Nebula are flashed our way 30 times a second.

Black holes

The cores of stars more massive than three Suns are crumpled up so thoroughly during a supernova explosion that nothing can stop the process of gravitational collapse. Its gravity becomes so great that nothing can escape from it—not even light. Black holes have effectively opted out of our Universe. The known laws of physics stop at the edge of a black hole. Some scientists think that black holes may provide "wormholes" in the fabric of space-time which allow vast distances within our own Universe to be traversed instantaneously. Black holes may even provide a gateway to alternative Universes.

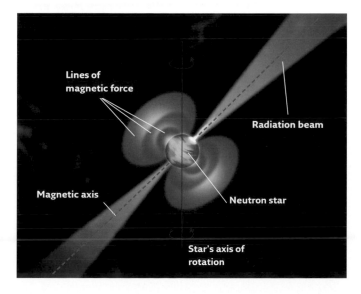

Lines of magnetic force

Radiation beam

Magnetic axis

Neutron star

Star's axis of rotation

Structure of a pulsar —the superdense, rapidly spinning remnant left after supernova explosions of the type that produced the Crab Nebula.

Multiple stars

Since we enjoy orbiting a single star, we are used to thinking of other stars as being singular objects. In fact, the Sun is in a minority of star-kind—most stars have partners, and some have more than one partner. But it is not always an equal partnership—the stars in multiple systems are often very different from each other, and show striking color contrasts.

A binary star comprises two stars in orbit about their common center of gravity. Many of the double stars visible through the telescope eyepiece are binary systems, but some are simply line-of-sight doubles—stars which are not gravitationally connected to each other, but which appear close to each other because of the effects of perspective.

Systems amazing

Castor, one of the stellar twins in the constellation of Gemini, was the first binary star to have been discovered. Its two main components—a blue A-type star and a red M-type star—orbit their common center of gravity in 350 years. Astronomers now know that Castor is an intriguing sextuple star system—each component of Castor is itself a binary star, while a faint red star farther afield makes another binary gravitationally bound to the system.

A number of lovely multiple stars are visible through small telescopes. One of the most well-known is the "Double-Double" star Epsilon Lyrae—a close pair of bright stars orbiting their common center of gravity, both resolvable as a double star. Perhaps the most perfectly framed multiple star is the celebrated "Trapezium"—a group of four stars within the Orion Nebula.

Colored double stars
Dozens of lovely colored double stars can be viewed through a small telescope, and some of the brightest and most beautiful examples are featured here. Colors have been exaggerated for clarity.

1. Beta Cygni (Albireo).
2. Eta Persei.
3. Epsilon Bootis.
4. Xi Bootis.
5. Gamma Delphini.
6. Iota Cassiopeiae.
7. Beta Orionis (Rigel).
8. Upsilon Andromedae.
9. Alpha Herculis.
10. Alpha Canum Venaticorum (Cor Caroli).
11. Alpha Scorpii (Antares).
12. Beta Scorpii.

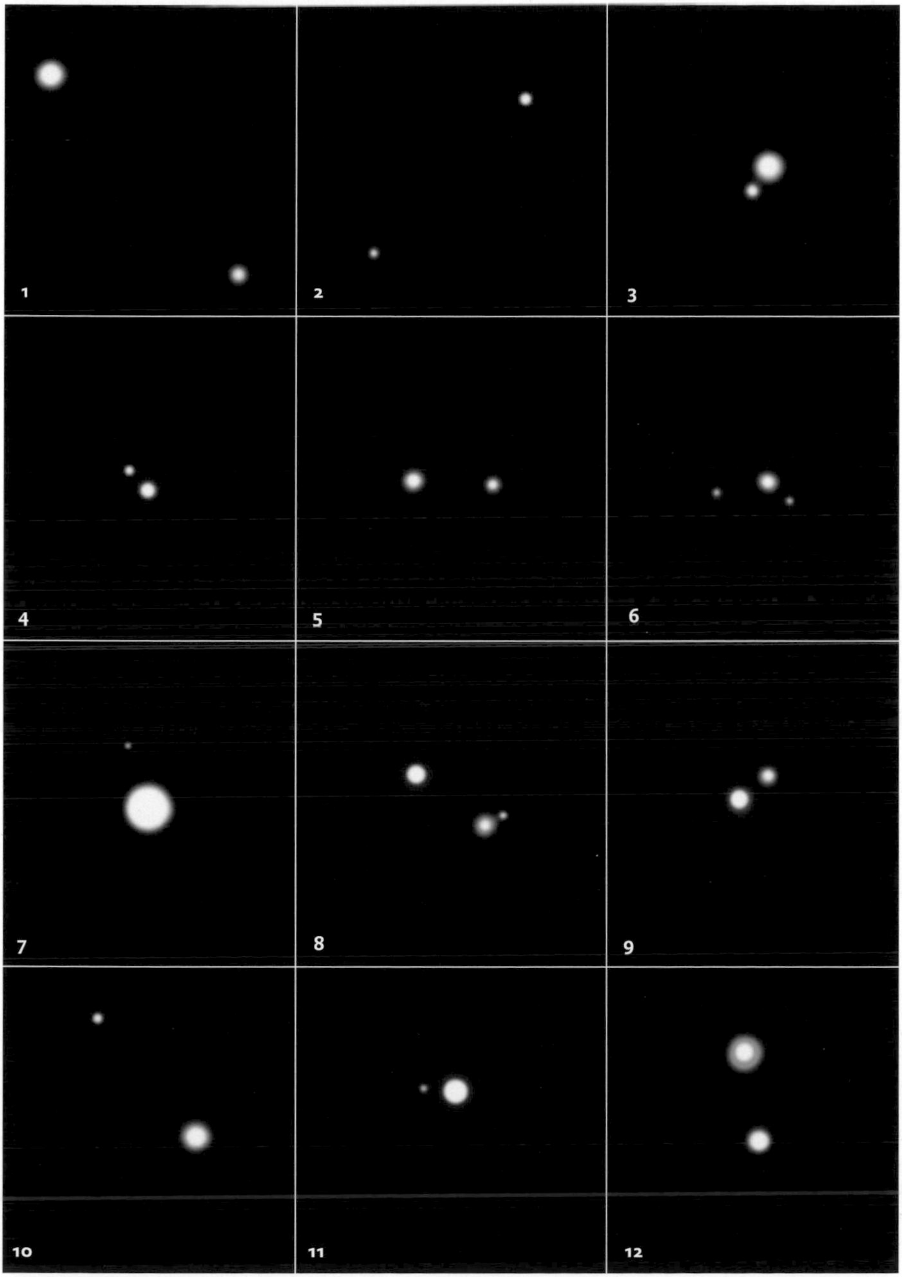
1
2
3
4
5
6
7
8
9
10
11
12

Variable stars

Everyone has noticed that bright stars sometimes twinkle, or even flash an array of colors. This is sometimes caused by unsteadiness in the Earth's atmosphere, or can also be a result of optical issues—refraction and reflection of light through the atmosphere—and has nothing to do with the output of light from the stars themselves. However, astronomers have discovered that many stars really do vary in brightness over a period of time, either on a regular, semi-regular, or irregular basis. Some stars show only small variations in their brightness, while others undergo major fluctuations, ranging over many orders of magnitude.

Celestial peek-a-boo

Known by its ancient Arabic name of Algol (the "demon"), the second brightest star in the constellation of Perseus experiences a substantial dip in brightness every 2.87 days, remaining at minimum for around ten hours. Algol is a type of variable known as an eclipsing binary. Its dip in brightness—easily noticeable with the unaided eye—is caused when it is eclipsed by a larger, dimmer star orbiting around it.

Another fascinating eclipsing binary is Epsilon Aurigae in the constellation of Auriga. Every 27.1 years, this yellow-white supergiant is eclipsed for two years by an unseen object that some astronomers think is a colossal star surrounded by a thick ring of dust. Its next eclipse is due between 2009 and 2011.

Cepheids

A class of yellow supergiants known as Cepheid variables vary in brightness as they rhythmically expand and contract over periods between two to 50 days. Cepheids are immensely valuable in determining the size of the Milky Way and the nearness of other galaxies, because their period of variability is directly related to their brilliance—the brighter the star, the longer its period. By comparing a Cepheid's apparent and actual brightness, its distance can be determined.

Algol, a famous variable star, is observed to fluctuate in brightness on a regular basis. The cause: a large, dim orbiting companion periodically eclipses Algol.

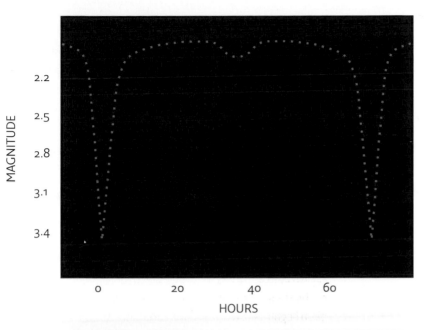

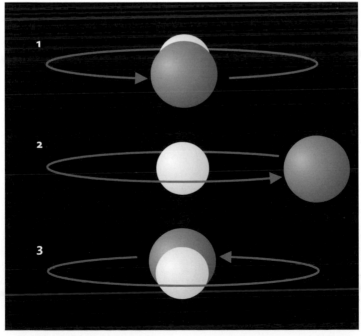

Long-period variables

Slowly pulsating long period variable stars expand and contract to produce changes in luminosity, just like Cepheids, but their cycles of variation only follow a general pattern. The best known long-period variable, Mira (the "wonderful") in the constellation of Cetus, has an average period of around 330 days during which it ranges from being easily visible with the unaided eye to being difficult to detect through binoculars.

Irregular variables

Stars with erratic changes in their brightness and/or period are known as irregular variables. There are two types—pulsating irregulars and eruptive irregulars. Pulsating irregulars are old supergiant stars approaching the end of their life, expanding and collapsing unpredictably as their nuclear fuel runs low. Eruptive irregulars undergo sudden increases in brightness as they eject material into space, their surfaces temporarily brightening to many times their previous magnitude.

Stellar cataclysms

Cataclysmic variable stars usually consist of a white dwarf primary and a red dwarf secondary. When a star in a closely orbiting binary star system leaves the Main Sequence towards the end of its life to become a white dwarf, its less massive red dwarf partner, still on the Main Sequence, finds itself orbiting a veritable stellar cannibal. The pair are so close together that the white dwarf's gravity distorts the shape of its neighbor, and its gaseous mantle of hydrogen is dragged away to form a hot accretion disk around the white dwarf. Occasional infalls of hydrogen gas onto the surface of the greedy white dwarf produce sudden outbursts called novae, so called because the star increases in brightness so much that it gives the appearance that a new star has suddenly appeared from nowhere (nova is Latin for "new").

Type I supernovae

Eventually, the white dwarf might accumulate so much mass that it implodes under its own gravity, leading to a supernova explosion and the creation of a neutron star. Because of the way they are produced, these kinds of stellar explosions, called Type Ia supernovae, are distinct from Type II supernovae which are described on page 130. Galaxies like our own are host to just a handful of supernovae each century.

Mira, a long period variable star, fluctuates in size and brightness over a period of around 330 days.

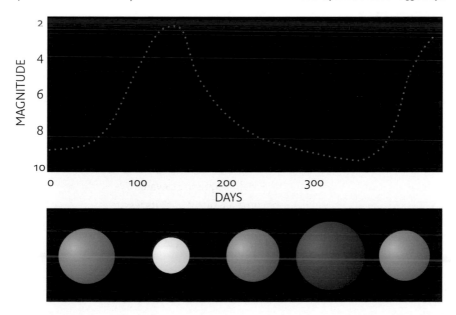

Stars in plane sight

From our position in the plane of a flattened spiral galaxy, the stars of the Milky Way recede into the distance, creating a glowing band that circles the sky.

Stars in their billions

100,000 light years in diameter, the Milky Way contains at least 200 billion stars. Our own Solar System lies around two-thirds of the way from the galactic center to its edge. Most of the stars in our vicinity formed from interstellar clouds of gas and dust within the galactic plane, as did the Sun, but a few appear to be just passing through—ancient visitors from the Milky Way's outer halo.

Single stars like the Sun are in a minority. Most stars belong to a multiple star system, in which two, three, or more stars are gravitationally bound together. Binary stars—two stars orbiting a common center of gravity—are the most common type of multiple star. Alpha Centauri is the best known example of a ternary (triple) star system (see opposite), in which a close binary is orbited by a more distant third star. There are other more complex star systems known which contain up to eight stars in stable orbits.

Twinkling near and far

Of all the stars in the galaxy, only 6,000 of them are bright enough to be seen with the unaided eye. So, from any given dark sky location on a clear, moonless night, those with good eyesight can discern around 2,000 stars at any one time—needless to say, nobody has ever succeeded in counting them all in one session of stargazing. All of the 6,000 stars that can be seen are fairly close to the Sun in galactic terms—the farthest, Iota-2 Scorpii (a supergiant

star in the constellation of Taurus) is 3,500 light years
distant, while light from the nearest, Alpha Centauri, takes
just 4.36 years to reach us.

Stellar neighbors

If you are at all familiar with science fiction then you will be
aware of the name Alpha Centauri. The nearest star system to
the Sun, Alpha Centauri, the brightest star visible in the
southern constellation of Centaurus, is actually a triple star
system consisting of the bright close double star Alpha Centauri
A and B, plus the much fainter red dwarf star Proxima Centauri,
which at 4.22 light years away has the distinction of being the
nearest star to the Sun, as its name suggests. Both Alpha
Centauri A and B are around the same size as the Sun, and orbit
their common center of gravity every 80 years. Alpha Centauri A
is remarkably like our own Sun, but just a little more massive
and slightly brighter, while Alpha Centauri B is a little cooler and
more orange in color.

Sizes of the Alpha
Centauri System
stars compared
with the Sun, and
their orbits.

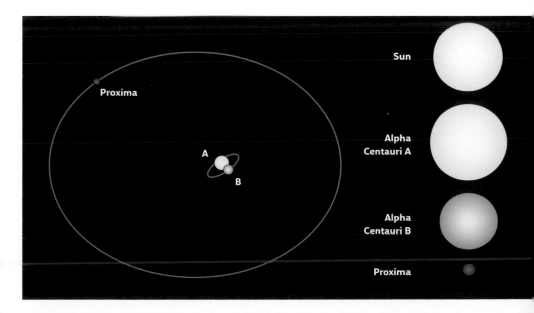

Stars in the 'hood

Although the Sun is a fairly bright middle-aged yellow G-type Main Sequence star, it is far from being an average kind of star, since the majority of stars in the galactic neighborhood are dim red dwarfs, too faint to be seen with the unaided eye. Only a handful of the 100 nearest stars shine with enough brilliance to be seen with the unaided eye. Out of the 100 brightest stars visible from the Earth, only three are nearer than ten light years, 35 are within 100 light years from us, and seven are more than 1,000 light years away.

Notable locals

Every type of Main Sequence star can be found in our cosmic neck of the woods, from faint M-type red dwarfs like Proxima Centauri to giant hot O-type blue stars like Zeta Puppis. There is a good proportion of brown dwarfs—objects more massive than planets but not quite massive enough to burn hydrogen at their cores and shine like stars. A fair sprinkling of white dwarfs—stars near the end of their lives—can also be found in our local galactic vicinity.

Just passing by

Barnard's Star, a dim red dwarf in the constellation of Ophiuchus, is just visible through binoculars. Lying six light years distant, it is the second closest star to the Sun after the Alpha Centauri system. In 1916, astronomer Edward Barnard discovered that the star has the largest known proper motion of all known stars—it moves against the celestial sphere at the staggering rate of 10.3 arcseconds per year, or

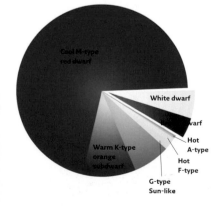

Cool M-type red dwarf

White dwarf

...warf

Hot A-type

Hot F-type

Warm K-type orange subdwarf

G-type Sun-like

Star types in the Sun's stellar neighborhood.

about half the apparent diameter of the Moon each century. Barnard's Star is approaching us at a relative velocity of 306,000 miles per hour, and in around 10,000 years' time it will make its closest approach to us, at a distance of 3.8 light-years.

Scorching Sirius

As well as being the brightest star in the constellation of Canis Major, Sirius is the brightest star in the entire sky—twice as bright as any other star. Located just 8.6 light years away, this brilliant A-type blue-white star is the fifth closest to the Sun. Of course, Sirius appears so bright because of its nearness to the Sun—there are many more intrinsically brighter stars. Sirius is partnered by a white dwarf, Sirius B, which orbits it every 50 years.

Stars which are located within ten light years of the Sun.

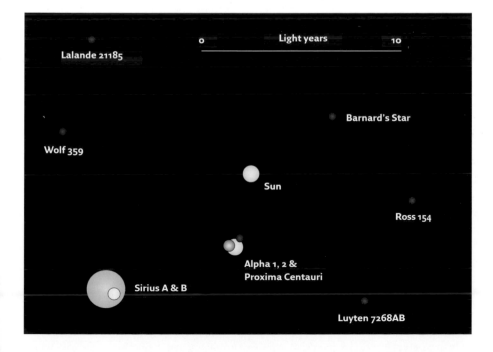

Open star clusters

More than 1,000 open star clusters are known within our own galactic vicinity. But since our view of the galaxy is largely obscured by dust and gas clouds in the plane of the Milky Way, this number likely represents just a small percentage of the true figure.

The beautiful "Jewel Box" open cluster lies in the Southern Cross. Its brightest stars shine a variety of colors.

Stellar siblings

Stars are rarely born in complete isolation. Our own Sun, now a singleton, was likely to have had a number of sister stars within the same galactic cloud of dust and gas from which it was born around five billion years ago. Clusters of newborn stars have been detected buried deep within galactic clouds of dust and gas, hidden from direct view but detectable by the heat that they emit in the form of infrared radiation.

Several million years after a galactic star cluster is born, its dusty, gaseous cocoon is dispersed by the strong stellar winds blowing from its hot new offspring. Over a period of a few hundreds of millions of years, the weakly gravitationally-bound stellar siblings become spread out along the galactic plane by wider gravitational forces.

Open star clusters offer astronomers valuable insights into the evolution of stars. Their constituent stars are all around the same distance from Earth, so their relative brightnesses can be compared without difficulty. Having been formed from the same cloud of dust and gas, stars within open clusters are all around the same age as each other and share similar chemical compositions. Despite this the stars within open clusters have a range of masses, from dwarfs smaller than the Sun to giants of up to a hundred times the mass of the Sun.

Seven sisters

Easily visible with the unaided eye, the sky's brightest open star cluster, the beautiful Pleiades in the constellation of Taurus, is a joy to behold through a pair of binoculars on a crisp, clear winter's evening. Around 380 light years distant, the hundreds of individual Pleiads are contained within a volume of space just 15 light years across. Estimated to be a mere 50 million years old, binoculars reveal a trace of nebulosity within the cluster—produced by fine dust reflecting starlight, long-exposure photographs reveal the extent of this hazy veil. Under good observing conditions, this nebulosity is especially noticeable surrounding the bright Pleiad star Merope.

The Pleiades star cluster and its surrounding reflection nebula.

Globular star clusters

Around 1500 hefty spherical star swarms called globular clusters, each consisting of hundreds of thousands of stars held together in a spherical mass by their mutual gravity, surround our galaxy in a vast halo.

Spectacular stellar spheres

Globular clusters are extremely ancient entities, their stars having been among the first formed in the galaxy. Since they have been discovered around many other galaxies, globular clusters are likely to be a common feature of most galaxies. An average globular contains hundreds of thousands of old red stars and measures more than 100 light years across. Stars become more tightly packed towards the center of a globular cluster, with the more crowded globulars having average star distances of only a few light weeks or months near at their cores.

Astronomers have attempted to understand how these vast spherical stellar congregations came into being, and how their globular shapes have been maintained for more than ten billion years. It is possible that giant black holes lurk at the centers of some globular clusters, providing the massive hub around which the cluster revolves. However, nothing so exotic exists within most globulars; instead, their densely packed stars undergo an uneasy gravitational waltz.

M13, the great globular cluster in Hercules.

Herculean cluster

M13, the largest globular cluster visible in northern skies, can be just located with the unaided eye in the constellation of Hercules. It is delightful to view telescopically, and many of its brighter stars can be resolved to its dense core through a 100mm telescope. It lies around 22,000 light years away and contains around 300,000 stars.

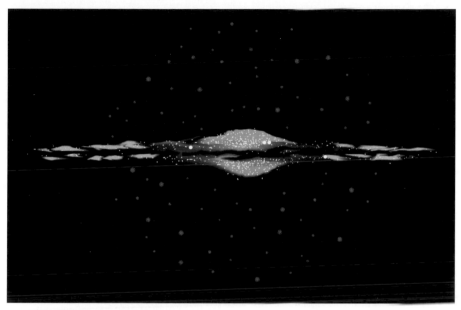

From our perspective within an outlying spiral arm of the Milky Way, globular clusters appear unevenly distributed around the sky, appearing more frequently in those constellations in and around the galactic center.

Globular clusters are distributed around the galaxy in a vast halo.

Giant globulars

In a league of its own, and by far the largest globular cluster in the vicinity of the Milky Way, Omega Centauri is a vast slightly flattened ball containing around ten million stars. Easily visible with the unaided eye as a large fuzzy patch in the southern constellation of Centaurus, Omega Centauri is some 150 light years across and lies at a distance of around 15,000 light years. A far larger giant globular cluster called G1 forms a miniature satellite galaxy to our near neighbor the Andromeda Galaxy. It is possible that such giants may be the core remnants of once larger galaxies whose outer stars have been gravitationally poached by their larger galactic neighbors.

Nebulae, dark and bright

The Milky Way contains an abundance of gas and dust clouds. Some are visible only because they appear in silhouette against a brighter starry background. Others shine by reflecting the light of nearby stars or by emitting light of their own.

The glorious Orion Nebula.

Dark clouds

Viewed under dark rural conditions, amazing detail can be seen along the band of the Milky Way. Much of the visible structure is produced by vast clouds of interstellar dust and gas along the plane of the galaxy which hide the light of distant starfields. A great dark rift running through the Milky Way in the constellation of Cygnus is produced by a dense galactic cloud several tens of thousands of light years long, separating our own spiral arm from the Cygnus arm of the galaxy. The southern constellation of Crux contains a smaller, though no less impressive dark cloud, known as the "coalsack"—a black oval covering around 30 square degrees, set against a bright stellar background. It is around 2,000 light years away and measures 60 light years across.

The mane attraction

Perhaps the most well-known dark nebula is the Horsehead Nebula in Orion—a remarkably shaped projection of the edge of a larger molecular cloud which is silhouetted against a glowing red gas cloud in the vicinity of the star Sigma Orionis. 1,500 light years away, the Horsehead Nebula is around ten light years long.

Hidden core

We can never see the center of our own Milky Way galaxy, in the constellation of Sagittarius, because it is obscured by intervening clouds of dust and gas. The bulge of old red stars around the galactic center rises above and below the galactic plane and is clearly visible on photographs. Observations made in wavelengths other than visible light allow astronomers to probe the heart of the galaxy. Infrared images have shown that a number of stars near the galactic core are orbiting a single compact object with the mass of a million Suns—an object very likely to be a supermassive black hole.

Shaped like a Staunton knight chesspiece, the lovely Horsehead Nebula, a column of dust and gas silhouetted against a glowing gas nebula background.

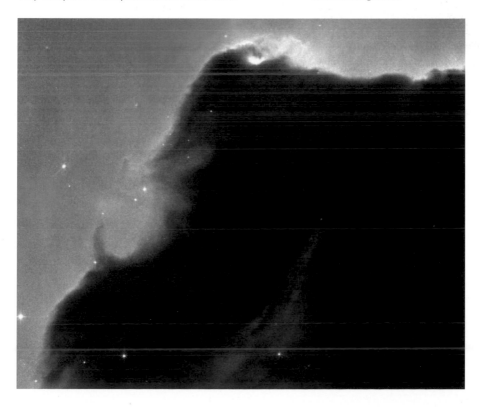

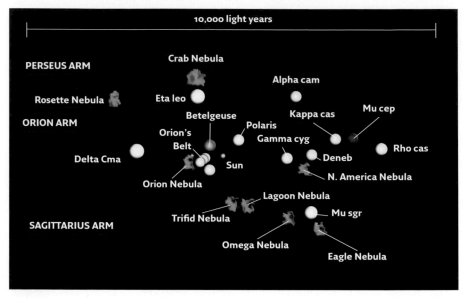

10,000 light years

PERSEUS ARM

Crab Nebula

Rosette Nebula

Eta leo

Alpha cam

ORION ARM

Betelgeuse

Kappa cas

Mu cep

Orion's
Belt

Polaris

Gamma cyg

Delta Cma

Sun

Deneb

Rho cas

Orion Nebula

N. America Nebula

Lagoon Nebula

Trifid Nebula

Mu sgr

SAGITTARIUS ARM

Omega Nebula

Eagle Nebula

The Sun's stellar and nebulous neighbors in the Milky Way, out to 5,000 light years.

On reflection

We have already seen how the light from bright newborn stars illuminates the surrounding dust and gas of their cosmic wombs, producing reflection nebulae. Their brightness depends upon the size and density of the grains of dust reflecting the starlight, and the color, brightness, and proximity of the stars which illuminate them. Reflection nebulae often have a pronounced blue color, caused by the reflective properties of carbon dust grains, blue wavelengths being more easily scattered than red light.

One of the most unusual of the sky's nebulae appears to fluctuate in brightness. Hubble's Variable Nebula in the constellation of Monoceros is a broad fan-shaped reflection nebula whose brightness changes with the variations of its illuminating star R Monocerotis, which lies at its apex. Streamers of dust near the star cast giant shadows onto the nebula's walls, producing an ever-changing spectacle which can be viewed through large amateur telescopes.

Emission nebulae

Intense ultraviolet light emitted by stars can make the hydrogen gas in their immediate neighborhood glow red. Known as HII regions, these emission nebulae can also be tinged with green hues, caused by ionized oxygen gas.

Of all the sky's emission nebulae, the Great Nebula in Orion is the brightest and most spectacular. Easily visible as a faint glowing patch in the sword handle asterism of Orion, a distinct green hue can be discerned through the eyepiece of an amateur-sized telescope. Photographs show a predominantly red nebula; delicate streamers of gas sweep gracefully away from the nebula's central region, which is punctuated by the four bright newborn stars of the Trapezium cluster. Nearby, a dark nebula intrudes against the brighter background glow, forming a feature known as the "shark's mouth." Observing it through his 1.2 meter reflecting telescope in 1789, William Herschel described it as "an unformed fiery mist, the chaotic material of future suns." Located some 1,600 light years away, the Orion Nebula is part of a far larger but much fainter diffuse nebula spanning virtually the whole of Orion.

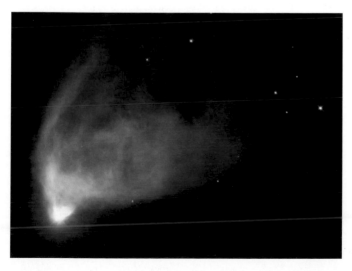

Hubble's Variable Nebula, a reflection nebula which varies in brightness with its illuminating star, R Monocerotis.

Galactic bubbles and froth

Since the Hubble Space Telescope began its incredible mission to image the Universe in 1990, it has revealed familiar deep space objects in spectacular detail.

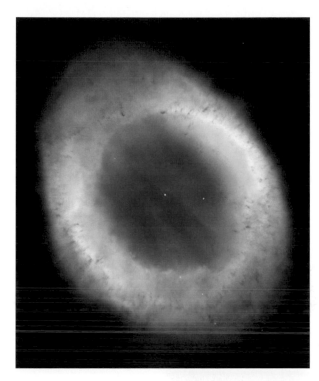

Beautiful detail is captured in this image of the Ring Nebula in Lyra. 2,300 light years distant, the nebula measures around a light year across.

Planetary nebula MyCn18, known as the Hourglass Nebula, is young by cosmic standards—its outer rings of gas were ejected just a few thousand years ago.

Barnard's Merope Nebula is a patch of reflection nebulosity just three light weeks from the bright star Merope in the Pleiades (just beyond upper right).

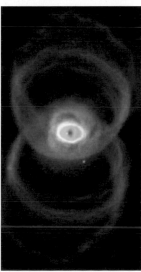

want to know more?

Take it to the next level...
▶ Birth of the first stars
  154
▶ Scale of our stellar
  neighborhood 19
▶ Scale of our galaxy 20
▶ Early notions about the
  Milky Way 58-9
▶ Cosmic distances
  measured 63

Other sources...
▶ Find out about our local
  galactic neighborhood
  with a good computer
  program.
▶ Some programs allow
  you to journey around
  space.

Weblinks...
▶ heritage.stsci.edu/
  gallery/gallery_
  category.html
  Fabulous deep space
  imagery on the Hubble
  Heritage image site.
▶ antwrp.gsfc.nasa.gov/
  apod/archivepix.html
  Astronomy Picture of
  the day features deep
  space images and
  explanations.

6 Far and away

Vast though our own Milky Way galaxy is, it is just one of billions of individual star systems scattered throughout the Universe. Some galaxies are far smaller, some appreciably larger than the Milky Way. Some have taken on graceful forms such as broad, curving spirals and smooth cigar-shaped ellipses, while others appear to have irregular shapes.

Peering into infinity

As we look into the depths of intergalactic space, we look at increasingly earlier epochs in the history of the Universe. The nearest galaxy is 180,000 light years away, while the light from the most distant known galaxies has taken more than 13 billion years to cross the Universe to meet us.

must know

Quasars
Exceedingly bright and distant objects called Quasars may be the result of galactic collisions in the early Universe, when close encounters between galaxies occurred more frequently than they do today. Galactic mergers added fuel to the accretion disks of supermassive black holes, producing exceptionally luminous regions of space.

Galactic evolution

By looking at galaxies over such a vast period of time, astronomers have gained an understanding of how galaxies were created and how they evolve. Galaxies were formed within large gas clouds very early in the history of the Universe. Exactly how the galaxies were seeded is not completely understood, but studies of the early Universe suggest that the gas within it—mainly hydrogen and helium— somehow developed ripples. In turn, knots of greater density collapsed under their own gravity to form the first star clusters and galaxies.

First stars

A greater abundance of hydrogen and helium gas in the early Universe meant that the first galaxies experienced high rates of star formation, producing predominantly hot blue stars. These massive stars, ranging from around 100 to 1,000 times the Sun's mass, burned out after several million years and blew themselves apart in huge supernova explosions, forming black holes. Heavier chemical elements produced within these exploded stars were scattered into their immediate cosmic environment, providing raw material more exotic than just hydrogen and helium for future stars and their planetary systems to assimilate as they formed.

Black hearted galaxies

Most galactic centers are thought to harbor a supermassive black hole with a mass ranging between a million and a billion times that of the Sun. As these immense black beasts pull matter toward them, they accumulate a ring of material known as an accretion disk. Frictional heating of the rapidly infalling matter produces a great deal of energy. Unsurprisingly, with such a great abundance of matter to devour, supermassive black holes at the cores of early galaxies are often highly active, incredibly bright, and energetic. Although our own galaxy's central black hole is currently quiescent, its gravitational effect can be seen on nearby stars.

Galaxies had a bumpy ride in the early Universe. Here, galactic collisions within the distant galactic cluster MS1054-03, 8 billion light years away, were imaged by the Hubble Space Telescope.

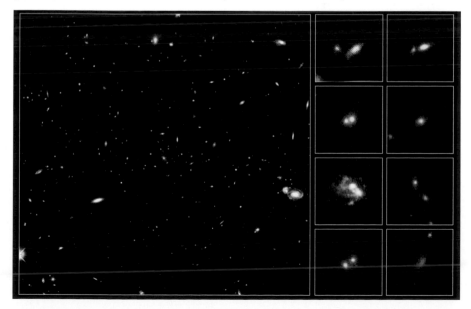

Dark matter
Just ten per cent of the matter within galaxies can be directly detected—a staggering 90 per cent of their matter is currently undetectable. Astronomers call this stuff "dark matter," and even though it produces no radiation or absorption, it is known to exist because its mass produces a measurable gravitational effect on galaxies. Some dark matter may be in the form of black holes, old neutron stars, brown dwarfs, and dust, but the majority of it remains unknown.

Galaxies of all shapes and sizes

Edwin Hubble devised a simple way to classify various galaxies by their shape—spiral, elliptical, and irregular. Spiral galaxies were noted to have two basic forms—the classic "grand design" spiral (S) like the Whirlpool Galaxy, and ones which had spiral arms emanating from a central bar of stars, called barred spirals (SB). The tightness of a spiral galaxy's arms was accounted for with a letter—for example, Sc is a loose spiral, while SBa is a tight, barred spiral. Elliptical galaxies were designated a number indicating variations in their shape, from 0 (circular) to 7 (highly elongated).

Spiral galaxies

Contained within a spiral galaxy's central bulge is a tightly-packed population of ancient, low-mass, red stars. In the surrounding flattened galactic disk, active star-forming regions illuminate gracefully sweeping spiral arms. Contained within these arms are brilliant young white and blue stars and their associated emission and reflection nebulae. At first glance, the nucleus and brightly glowing arms of a spiral galaxy appear to constitute its entire fabric, but in fact, the dark lanes that separate a galaxy's bright spiral arms are opaque clouds of dust and gas—the material from which populations of future stars will be born.

Spiral spin

Measurements of the rotation speeds of spiral galaxies reveal that stars in the outer reaches of their spiral arms revolve around the galactic center at much the same velocity as those stars nearer the center. This might appear to violate Kepler's orbital laws, which state that the velocity of orbiting objects decreases with distance from the gravitational hub—be they planets orbiting stars, or stars orbiting the galactic center. However, the situation will be different if the entire galaxy were embedded within a halo of dark matter with many times the galaxy's mass. Then, the gravitational influence of the dark matter would be sufficient to balance things out, producing a more constant galactic spin.

Barred galaxies

Around one in three spiral galaxies contain a prominent elongated bar of bright, mainly yellow stars

The barred spiral galaxy NGC 4319 in the constellation of Draco. To the upper left of the galaxy is the bright but more distant quasar Markarian 205.

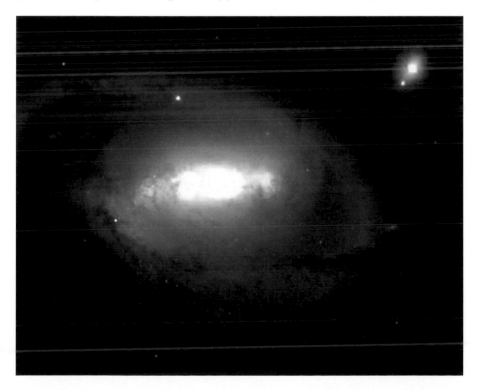

across their centers which can stretch the galaxy's overall visible diameter up to fifty per cent. Spiral arms sprout from either end of the bar, with the spiral arm and bar often making a relatively sudden, sharp angle with each other. There are indications that our own Milky Way galaxy may be slightly barred, rather than having arms which curve all the way into its central bulge.

Barred spirals certainly make a rather peculiar sight, and the mechanics of their formation is currently not very well understood. It seems likely that galactic bars are temporary features on the cosmic timescale. Their formation may be triggered by gravitational instability near a spiral galaxy's core, throwing the tightly packed stars in the galactic hub into elongated orbits. Alternatively, bars may be produced by gravitational disruptions caused by nearby galaxies.

Elliptical galaxies

Around one in ten of all galaxies is a flattened football-shaped agglomeration of stars known as an elliptical galaxy. They range from dwarf ellipticals a few light years across and with the mass of ten suns, to some of the biggest galaxies in the Universe— hefty giant ellipticals up to half a million light years in diameter with the mass of several trillion Suns. Elliptical galaxies contain fewer young stars and far less gas and dust than spiral galaxies, and their stars appear to move in a more random manner than those of spiral galaxies.

Elliptical galaxies display little structure, apart from their elliptical shape and a denser concentration of stars toward their center. Hubble's classification of elliptical galaxies indicates the degree to which they appear non-spherical—a near-spherical elliptical galaxy is classed E0. However, appearances can be deceptive; while a highly elliptical galaxy is classed E7 if it happens to be viewed from its side, a similar elongated galaxy viewed end-on will appear foreshortened and spherical, and will be classed E0.

NGC 205, a dwarf elliptical companion to the nearby Andromeda Galaxy.

When swirls collide

Although astronomers of the early-20th century were inclined to call galaxies "island Universes," early 21st century astronomers know that no galaxy is an island— each is subject to the gravitational forces at play within its environment, from the vast cloud of dark matter in which it is embedded to the influence of nearby galaxies.

Close galactic encounters and collisions have played a major part in the evolution of many galaxies, leading to widespread structural changes and bursts of new star formation. Tidal interactions occur when two galaxies pass near each other, without their visible matter—dust, gas clouds, and stars—actually mixing. Such close galactic encounters produce distortions in their shapes, including warping along the planes of spirals, tidal bulges, and streams of matter.

Colliding spiral galaxies NGC 2207 and IC 2163 in the constellation Canis Major, 114 million light years distant. The smaller IC 2163 is around the same size as the Milky Way; in due course it will be assimilated into its larger neighbor.

When galaxies collide, they do so at a pace that cannot be directly observed—the average collision speed between two medium-sized galaxies being around 300 km per second. Computer simulations, however, have provided an excellent means of observing the behavior of a variety of different galaxies colliding at different velocities and angles. Although galaxies can intersect and mingle intimately with each other, their individual stars are so widely spaced that they rarely collide with one another. Gas clouds within colliding galaxies, however, invariably slam into each other, producing waves of compression that create conditions suitable for the formation of new stars.

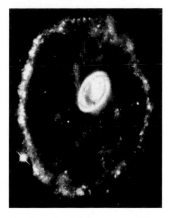

The Cartwheel Galaxy, 500 million light years away in the constellation of Sculptor—the result of a head-on collision between galaxies.

Pulling on the pool

One of the best and brightest examples of an interacting galactic pair is the Whirlpool Galaxy and its smaller neighbor NGC 5195, around 30 million light years distant in the constellation of Canes Venatici. Connected by a bridge of dust, gas, and young stars, the Whirlpool's beautiful spiral structure has been largely produced by the close passage of NGC 5195, a smaller irregular galaxy.

Head-on collision

In rare instances, some galaxies can pass right through each other. The Cartwheel Galaxy, 500 million years away in the constellation of Sculptor, obtained its remarkable shape when two galaxies experienced such a head-on collision. Ripples of energy sent 184,000-mph compression waves into the surrounding gas and dust, producing a burst of star formation in a vast ring around the impacted galaxy.

Local heroes

Our own quaintly-named local group of galaxies contains no fewer than 30 galaxies of varying types and sizes, contained within a sphere some ten million light years across.

Magellan's clouds

Named after the 16th century seafarer Ferdinand Magellan, the Small Magellanic Cloud (SMC) and Large Magellanic Cloud (LMC) are the Milky Way's two largest satellite galaxies. Both are located deep in southern skies and are easily visible without optical aid as sizeable smudges about 20° apart, looking like detached portions of the Milky Way. The SMC occupies a corner of the constellation of Tucana, and measures around 3° across; the LMC appears about twice as large and spans the border between the constellations of Dorado and Mensa.

More than 200,000 light years away, the SMC is a dwarf irregular galaxy containing around 30 billion stars. Its diffuse appearance in visible light, and the scattered nature of its various centers of activity seen in other wavelengths, means that it is not possible to pinpoint a center to the SMC, except that its stars are more concentrated toward an elongated axis several thousand light years long. Indeed, the SMC may once have been a barred spiral galaxy that was gravitationally disrupted by the Milky Way.

Around 160,000 light years distant, the LMC is a dwarf irregular galaxy about five percent the diameter of the Milky Way, containing around ten billion stars. Like the SMC, there are indications that the LMC was once a small barred spiral galaxy which has been gravitationally disrupted as it passed close to the Milky Way—a distinct bar of stars within the LMC is still visible. Deep sky delights abound within the LMC, including the brightly glowing spidery form of the Tarantula Nebula, its center bustling with dozens of bright supergiant stars. In and around the LMC are numerous glorious open star clusters and nebulae. In February 1987, a bright supernova erupted within the LMC, becoming the first supernova visible to the unaided eye since Kepler's supernova of 1604. Known as SN 1987A, the supernova gradually dimmed in the months following its appearance, and powerful telescopes can detect glowing rings surrounding the supernova, illuminated by energy from the blast. The remnant's central parts are brightening as material thrown out by the supernova slams into previously ejected gas.

The Small Magellanic Cloud, a dwarf irregular satellite galaxy which orbits the Milky Way.

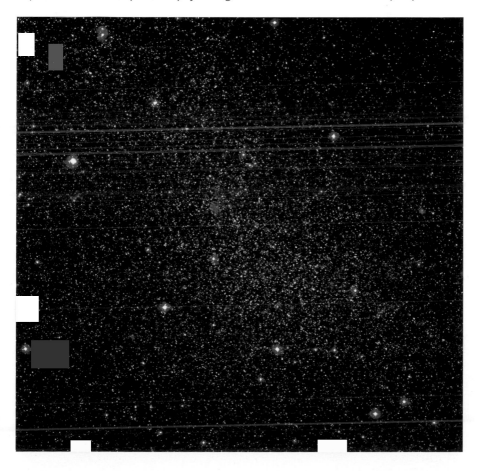

The Milky Way's dwarf satellites

In addition to the SMC and LMC, the Milky Way has a number of dwarf satellite galaxies, many of which were discovered only in recent years because of their faintness and the apparent proximity of some to the crowded galactic plane.

Canis Major—CMa DIrrG

A dwarf irregular galaxy only 40,000 light years distant, discovered in 2003. The nearest galaxy to the Milky Way, it contains around a billion stars, most of which are red giants. Several globular clusters are associated with this galaxy, which is experiencing severe disruption by the Milky Way's gravity.

Sagittarius—Sgr DSphG

A dwarf spheroidal galaxy around 50,000 light years distant, 10,000 light years in diameter, discovered in 1994. This sizeable galaxy is nestled on the far southern side of our galaxy and difficult to observe because of intervening material. It orbits the Milky Way every billion years, and during this close passage it is being torn to shreds by our galaxy's gravity.

The second nearest galaxy to our own, the Sagittarius dwarf spheroidal galaxy is so transparent that distant galaxies can be clearly glimpsed through it.

Ursa Minor—UMi DSphG

A dwarf spheroidal galaxy around 1,000 light years across and 220,000 light years away, discovered in 1954. It contains almost exclusively ancient red stars.

Sculptor—Scl DSphG

A dwarf spheroidal galaxy around 1,000 light years across and 260,000 light years away, discovered in 1938.

Draco—Dra DSphG

A small dwarf spherical galaxy around 500 light years across and 270,000 light years away, discovered in 1954. Globular cluster M54 is thought to belong to this small, ill-fated galaxy.

Sextans—Sex DSphG
A dwarf spherical galaxy around 3,000 light years across and 290,000 light years away, discovered in 1990.

Carina—Car DSphG
A small dwarf spherical galaxy around 500 light years across and 330,000 light years away, discovered in 1977. Its stars appear to be several billion years younger than the other small dwarf spheroidal galaxies.

Ursa Major—UMa DSphG
A dwarf spheroidal galaxy around 3,000 light years across and 330,000 light years away, only discovered in 2005. It is the least luminous galaxy known.

Fornax—For DSphG
A dwarf spheroidal galaxy around 3,000 light years across and 450,000 light years away, discovered in 1938.

Leo—Leo II DSphG
A small dwarf spheroidal galaxy around 500 light years across and 670,000 light years away, discovered in 1950. This tiny galaxy of several million stars has a number of orbiting globular clusters.

Leo—Leo I DSphG
A dwarf spheroidal galaxy around 1,000 light years across and 830,000 light years away, discovered in 1950.

The Milky Way's most distant satellites are the **Phoenix Dwarf Galaxy**, 1.2 million light years distant and discovered in 1976, and **NGC 6822**, 1.8 million light years away and discovered visually by Edward Barnard in 1884 using a 150mm refractor. Both are small, irregular galaxies.

The local group

More than thirty galaxies of various sizes make up the local group, a cluster of galaxies dominated by the big three—our own Milky Way, the Andromeda Galaxy, and the Triangulum Galaxy—around which most of the smaller galaxies are huddled.

The great Andromeda spiral

2.9 million light years distant, the Andromeda Galaxy is the farthest thing visible with the average unaided eye. A majestic spiral galaxy with a diameter of about 250,000 light years—twice the size of our own Milky Way—it appears tilted to us about 30°

To the unaided eye, the Andromeda Galaxy appears as a faint elongated smudge.

from its galactic plane, so we see only a rather foreshortened view. It would present quite a sight if it were face-on to us. Interlaced with its bright starry spiral arms are bright and dark nebulae—considerable structure can even be made out through backyard telescopes. Two bright dwarf elliptical satellite galaxies—M32 and M110—lie near the Andromeda Galaxy.

Approaching us at a velocity of 306,000 mph, the Andromeda Galaxy is set on a collision course with the Milky Way. A merger between the two galaxies will take place in about three billion years' time, and computer models suggest that a large elliptical galaxy will be produced as a result. Galactic mergers appear to be a phenomenon that the Andromeda Galaxy is used to—astronomers have discovered that it has a double nucleus, indicating that it has assimilated a smaller galaxy which once passed by its core, leaving only the intruder's nucleus as evidence of the ancient merger.

Pinwheel in Triangulum

The smallest spiral in the local group, the Pinwheel Galaxy in the constellation of Triangulum spans some 50,000 light years. It lies at a distance of more than 3 million light years from us, but only 700,000 light years from the Andromeda Galaxy. It appears more face-on than the Andromeda Galaxy, but its low surface brightness means that it is not as spectacular a telescopic sight. High resolution photographs show it in all its glory—a loosely wound anticlockwise spiral, with blue knotty arms and a bright reddish nucleus.

Galactic clusters

Galactic clusters are classed as regular or irregular, depending on certain characteristics. Regular clusters have a marked symmetry, with a greater number of galaxies near their centers. Irregular clusters have little symmetry to their shapes and the galaxies are fairly evenly spread about.

Farthest galaxy
A galaxy called Abell 1835 IR1916, discovered in 2004, is about 13,230 million light-years away. We see it just 470 million years after the Big Bang, when the Universe was just three percent its current age.

Virgo cluster

Although our own local group contains more than thirty individual galaxies, it is rather small compared to most galactic groupings we observe. Around 50 million light years distant, in the direction of the constellation of Virgo, lies a far larger cluster of galaxies. A hundred times larger than our own local group, the Virgo cluster contains around 3,000 individual galaxies, all held together in space by their mutual gravitational attraction. The Virgo cluster measures around ten million light years across, and from our vantage point within the Milky Way galaxy, the part of the sky occupied by it can be easily covered by an outstretched hand. One in every hundred of the Virgo cluster's galaxies is a large spiral. Its largest member, the giant elliptical galaxy M87 at the cluster's core, is one of the biggest galaxies in the Universe, and contains several trillion stars. So great is the Virgo cluster's gravity that it forms the physical center of the galactic supercluster to which we belong. Our local group will probably be pulled toward the Virgo cluster and become absorbed into it in the far future.

More than ten times the distance of the Virgo cluster, at a distance of around 650 million light years in the constellation of Hercules, lies another major galactic cluster. An estimated 100,000 galaxies populate the crowded Hercules cluster—many of them appear to be interacting or colliding with their neighboring galaxies.

Superclusters

Professional telescopes are becoming larger and more powerful, and their sensitivity to faint light from the depths of the Universe increases. As we look farther across the unimaginably vast Universe, peering back through time, we gain a deeper understanding about its large-scale structure. We have discovered that galactic clusters congregate into superclusters which are distributed throughout the cosmos in an intricate latticework of strings, sheets, and walls surrounding enormous voids.

Bright nearby galaxies M81 and M82 in Ursa Major—the fainter galaxy NGC 3077 can be seen at lower right.

Hubble's galactic showcase

In surveying the depths of the Universe, from the nearby Magellanic Clouds to the faintest celestial smudges at the edge of time and space, the Hubble Space Telescope has revealed that the cosmos is more fantastic than we can possibly imagine.

NGC 3370 is a lovely grand-design spiral galaxy in the constellation of Leo. 100 million light years distant, it is about the same size as our own Milky Way. A number of distant galaxies can be seen in the background.

The Hubble Space Telescope captured this glorious image of the barred spiral galaxy NGC 1300 in the constellation Eridanus. The galaxy is around the same size as our own Milky Way and is 70 million light years distant.

A jet of electrons and subatomic particles shoots out at near light speed from the center of the giant elliptical galaxy M87. Twisted magnetic fields generated by superheated gas falling into a supermassive black hole at the galaxy's core are thought to power the jet. M87 is around 50 million light years away.

In 2003 the Hubble Space Telescope secured our deepest look into the Universe by spending three months looking at a single small spot of the sky. The Hubble Ultra Deep Field shows a multitude of galaxies as they appeared around 13 billion years ago.

want to know more?

Take it to the next level...
▶ Scale of our local galactic environment 20
▶ Structure of the wider cosmos 21
▶ View of our galaxy and its globular halo 145
▶ How telescopes reveal the Universe in a range of wavelengths 174–81

Other sources...
▶ Binoculars will easily reveal several of the brighter nearby galaxies.
▶ *Need to Know? Stargazing* will help you locate brighter nebulae and galaxies.
▶ Explore the galaxies with a good computer program.

Weblinks...
▶ www.seds.org/messier/galaxy.html
A guide to galaxy types, with lots of detailed information.
▶ chandra.harvard.edu/index.html
Galaxies galore revealed by NASA's orbiting Chandra X-ray observatory.
▶ www.nasa.gov/vision/universe/starsgalaxies/
NASA's guide to stars and galaxies.

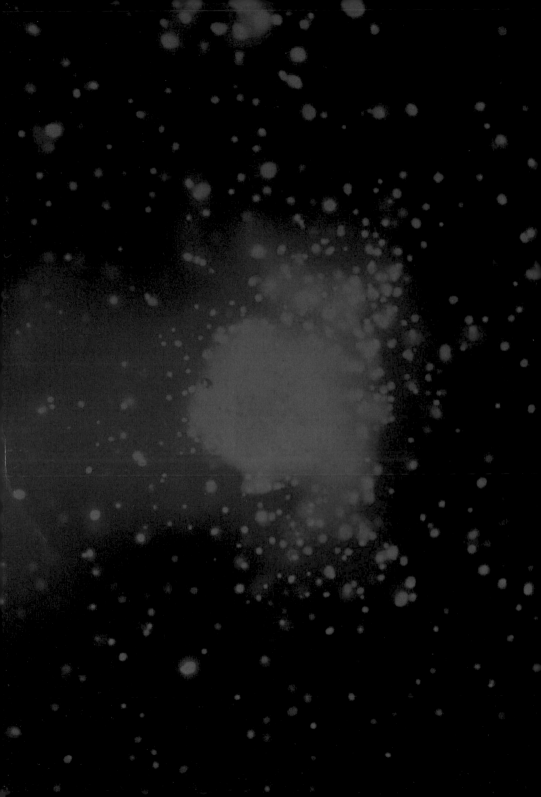

7 The Universe revealed

For nearly 400 years telescopes have peered into the depths of the Universe. During the last 150 years, photography and spectroscopy have enabled light to be captured and analyzed, and a host of wavelengths invisible to our eyes— such as infrared, radio, and gamma rays—have helped build a picture of the processes that are at work in the Universe.

Gathering light

Optical telescopes collect and focus ordinary light, providing bright, magnified views of all kinds of objects in the Universe, from the nearby Moon to distant galaxies.

Scope for discovery

Early astronomical discoveries were made under chilly night skies by enthusiastic amateur astronomers using home-made telescopes slung on shaky mounts. Now professional astronomers operate huge state-of-the-art telescopes (on the Earth and in Earth orbit) by pushing buttons from the comfort of a warm control room. Discoveries about the Universe continue to be made, and they are being made at an ever-increasing pace. Some new insights into the Universe are rather mundane, like finding a new Cepheid variable star in a distant galaxy; and some of them significant, like the discovery in 2005 of a distant planet in our Solar System larger than Pluto.

The Isaac Newton Telescope peers into the Universe from a mountain summit on La Palma, high above the clouds.

Amateur astronomy

Astronomy is one of the few sciences where non-professionals can make significant contributions and new discoveries. Encouragingly, the gap between amateur and professional astronomers is getting smaller, thanks to the increasing quality and sophistication of commercially-available optical equipment and accessories. Thanks to computers and light-sensitive CCD chips, amateur astronomers with comparatively modest means are capable of securing images that equal those taken by large professional observatories several decades ago. Amateur telescopes with go-to technology can rapidly locate tens of thousands of celestial objects at the push of a keypad button.

But technology is not the only leveler. Amateur astronomers have the advantage of possessing a deep love for their subject and being able to devote as much time as they want to their chosen specialty. Amateurs can spend time observing objects infrequently visited by the professionals, such as the Moon and planets, monitoring variable stars— and observing phenomena rarely witnessed in a professional capacity, such as meteors.

A spectrum of possibilities

Our eyes are capable of seeing only visible light, which is just a small part of a vast range of wavelengths of light known as the electromagnetic spectrum. By looking at objects in a variety of wavelengths across the electromagnetic spectrum—from long wavelength radio frequencies to short wavelength gamma rays— astronomers can learn about the energy produced by processes taking place within them, on their surfaces or in their atmospheres and immediate environments. All radiation, across the entire electromagnetic spectrum, travels at the speed of light.

Ever-resourceful, amateur astronomers around the world have organized themselves into groups dedicated to pursuing interesting lines of observation and imagery with the potential to make major discoveries. Amateurs frequently discover and monitor changes on the planets; they often find new comets, and amateur searches have brought to light many supernovae in distant galaxies.

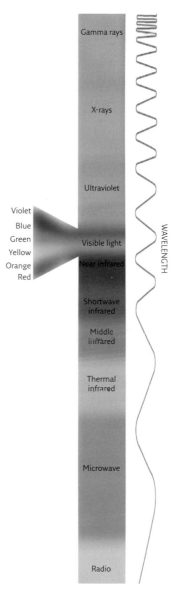

Astronomers observe along all bands of the electromagnetic spectrum, from Gamma rays, through visible light to radio waves.

Tuning in to the cosmos

The days when professional astronomers placed their eyes to a telescope and peered at celestial objects disappeared with the arrival of photography and spectroscopy in the late 19th century. Astronomers of the early 21st century observe in new wavelengths.

The view from Earth

Telescopic observation from the Earth's surface is hampered by the atmosphere. Turbulence in the atmosphere causes visible light to shimmer, degrading the quality of the images produced by optical telescopes. From low-lying locations beneath most of the Earth's troposphere (the lowest, densest part of the atmosphere), only visible light and radio wavelengths are capable of penetrating the atmosphere to be observed. Atop a high mountain, where the atmosphere is thinner and there is less water vapor, infrared wavelengths are capable of being detected from cosmic sources. Shorter wavelengths than visible light are absorbed by the upper atmosphere. To study objects in ultraviolet, X-ray, and gamma ray wavelengths, our observatory needs to be in space, high above the Earth's atmosphere.

Visible light

Celestial objects appear to shine either because they are producing light of their own or they are reflecting light from nearby bright objects. Stars are intrinsically bright, but the planets around them are dark objects, visible only because they are illuminated by the light of their parent stars.

Infrared

Every object in the Universe (with the exception of black holes) emits infrared radiation, or heat energy. Infrared observatories have given astronomers unprecedented views of areas previously hidden from view by dust clouds within the Milky Way and other galaxies, because infrared radiation can pass unhindered through these clouds while visible light is blocked and scattered. Planet-forming disks have been detected around other stars, and brown dwarfs—objects with too little mass to have developed into true stars—have been detected in many nebulae and star clusters. It is possible to observe distant galaxies in the infrared because their light has been highly redshifted by their rapid recession.

Microwave astronomy

Observations of the cosmic microwave background radiation — the fossilized glow from the Big Bang—show that 379,000 years after the Big Bang, the Universe had become clumpy, sowing the seeds of galaxy formation.

NASA's Spitzer Space Telescope imaged the Ring Nebula in infrared light, revealing that the familiar bright ring is surrounded by delicate petals of faintly glowing gas.

Radio astronomy

Radio waves are emitted by a variety of celestial objects and phenomena, from the Sun to distant galaxies. Astronomers use large metallic dish antennae to collect and focus these radio signals. Images are built up electronically, based on the strength of the received signals over a portion of the sky, so like their optical counterparts, radio telescopes usually have a steering mechanism to keep them pointed in the same direction as the Earth rotates. Because radio waves have much longer wavelengths than visible light, the collecting area must be quite large—individual professional radio telescopes have diameters ranging from a few meters to many tens of meters.

Little green men

Since radio telescopes came into widespread use by astronomers in the late 20th century, they have made a remarkable contribution to our knowledge of the Universe. One of the most astonishing discoveries took place in August 1967 when Antony Hewish and Jocelyn Bell-Burnell stumbled across a cosmic radio signal unlike any other previously detected—it pulsed every 1.337 seconds, without missing a beat, keeping

Searching for extraterrestrial activity, the SETI@home project uses the free computing time of millions of home PCs to analyze radio signals from deep space.

perfect time. Initially it was uncertain whether the pulsing signal came from an artificial or natural source—the signal was even jokingly given the label LGM-1, for Little Green Men 1, just in case it happened to be produced by an alien civilization! It was eventually identified as a pulsar—a rapidly spinning neutron star emitting a narrow beam of radio waves— the ultradense product of a supernova explosion. Many more pulsars have since been discovered, some of them associated with visible supernova remnants, most famously the Crab Nebula and its pulsar.

More little green men

One way in which an advanced extraterrestrial civilisation might be detected is by listening for any obviously artificial radio signals coming from deep space. Random radio noise pervades the sky, and it has only been possible to analyse signals collected by radio telescopes in detail in recent years because of advances in computing power. A novel approach to analysing mountains of radio data was commenced in 1999 as the SETI@home project, run by the University of California, Berkeley. Users of home computers signing up to the project download a program that processes data while the machine is idle.

In 1974 a signal was broadcast into deep space by the Arecibo radio telescope, in the hope that some attentive alien civilization might pick up the message—and be able to decode it.

Binary numbers 1–10

Atomic numbers of essential elements

Chemical formulae for molecules important for life

DNA molecule double helix and a human chromosone

World population, a human(1) and the height of humans

Solar systems

Arecibo radio telescope

Size of telescope

The ultraviolet Universe

Ultraviolet light observations allow astronomers to discover the temperature and composition of a range of objects in the Universe—from cold interstellar gas and dust clouds to hot young stars. Ultraviolet observations have provided information about our galactic neighborhood as well as more distant galaxies.

X-rays and gamma rays

X-rays and gamma rays are produced by a variety of exotic, energetic objects and phenomena in the Universe, including white dwarfs, supernovae, neutron stars, and pulsars, black holes and active starburst galaxies. Europe's XMM-Newton has observed cosmic X-ray sources since 1999. One of its many discoveries was the most distant galaxy cluster in the Universe, formed when the Universe was just 5 billion years old. NASA's orbiting Chandra X-ray observatory is providing

The glowing supernova remnant Cassiopeia A, seen in a composite of images taken in three different wavelengths by NASA's Spitzer, Hubble, and Chandra space telescopes.

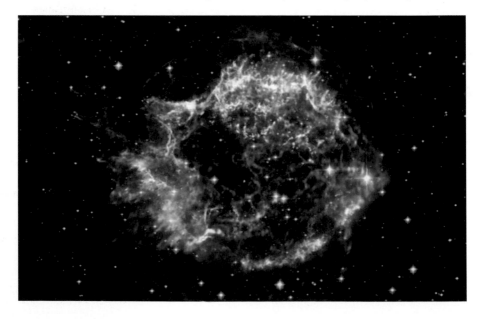

astonishingly detailed views of the Universe, from nearby Solar System objects to X-ray sources many billions of light years distant.

Prospectus Universalis

Despite the increasingly-polluted atmosphere of our home planet, ground-based astronomy—both amateur and professional—will continue to thrive with advances in optics, engineering, and technology. The amazing success of orbiting observatories like NASA's Hubble Space Telescope and NASA's Chandra X-ray telescope has paved the way for bigger, better instruments. Across the entire electromagnetic spectrum, unprecedented detail will be revealed in known objects, and a host of strange new objects will emerge as our cosmic eyes widen and adapt to the darkness.

Planets will continue to be discovered around distant stars—it won't be long before an Earth-like world is found orbiting a Sun-like star. It is not fanciful to imagine that an advanced telescope will be able to image such a planet's continents and oceans and detect the signature of life in the spectrum of its feeble light.

Meanwhile, spaceprobes will continue to explore the Solar System. Softlanders, rovers, gliders, and balloons will nose around strange landscapes. Within the next couple of decades, humans are likely to once more set foot upon the Moon, after an absence of around half a century, and the orange soil of Mars will be trodden upon soon afterward. Perhaps primitive life will be found elsewhere in the Solar System—such a discovery would indicate that extraterrestrial intelligence is not such a rare thing in the cosmos. Perhaps the rarest thing is a species intelligent and curious enough to look out into the Universe—to wonder, imagine, and explore.

want to know more?

Take it to the next level...
▶ **Ancient cosmological notions and theories** 28–37
▶ **How Galileo saw the Universe with one of the first telescopes** 41–2
▶ **The most distant galaxies** 168–9

Other sources...
▶ *Need to Know? Stargazing* will help you discover the Universe with your own eyes.
▶ Tune out—some of the static fuzz on an untuned TV screen is caused by the cosmic microwave background radiation—the echo of the Big Bang!
▶ Join a local or national astronomical society.

Weblinks...
▶ planetquest.jpl.nasa. gov/index.cfm
▶ PlanetQuest—NASA's search for planets around other stars.
▶ www.jwst.nasa.gov NASA's James Webb Space Telescope, under development.
▶ www.eso.org/outreach /ut1fl/ VLT—the European Very Large Telescope, atop a mountain in Chile.

Glossary

Accretion
The gathering together of material. On a small scale, accretion works when two tiny particles collide and stick together. On a large scale, the gravity of individual particles attracts other particles, producing gradual growth of particle size. Accretion formed the planets and their satellites, asteroids, and comets. Many black holes are surrounded by accretion disks—matter that they have attracted from their immediate environment.

Asteroid
A large chunk of rock orbiting the Sun, ranging from a few tens of meters to a few hundred kilometers in diameter. Also called a minor planet.

Asteroid belt
A zone of the Solar System containing a large number of asteroids. The main asteroid belt lies between the orbits of Mars and Jupiter.

Astronomical Unit (AU)
The average distance from the Earth to the Sun —around 92 million miles.

Astronomy
The science-based study of celestial objects and phenomena.

Big Bang
The colossal explosion at the beginning of time that created the Universe and everything in it. It is thought to have taken place around 13.7 billion years ago.

Black hole
A relatively small region of collapsed space-time with such a huge gravity field that nothing—not even light itself—can escape.

CCD
A charge-coupled device, a light sensitive electronic chip used in astrophotography.

Comet
A city-sized chunk of ice and rock which heats up on entering the inner Solar System, emitting gas and dust which forms a coma, and perhaps developing a long tail of gas and dust.

Constellation
A precisely defined region of the sky created to enhance our familiarity with the heavens. There are 88 recognized constellations, some of which date back to antiquity.

Core

The central region of a star or large planet—usually very hot and under extreme pressure. Many minor planets are chips from larger objects and may not have a well-defined core.

Dark matter

A currently undetectable form of matter which makes up 90 percent of the mass of the Universe, known to exist because its gravity influences the motion of galaxies.

Double star

A pair of stars which appear close together in the sky. Some are binary systems orbiting each other; others, produced by line-of-sight perspective, are known as optical doubles. Systems of three or more stars are called multiple stars.

Electromagnetic radiation

All energy in the electromagnetic spectrum—from short wavelength gamma rays to long wavelength radio waves—is propagated through space at the speed of light by vibrating electrical and magnetic disturbances. Visible light is a form of electromagnetic radiation.

Exoplanet

A planet in orbit around a distant star. Hundreds are currently known.

Galactic cluster

A group of galaxies held together by their mutual gravitational attraction. Galactic clusters themselves belong to even larger superclusters —the largest structures in the Universe.

Galaxy

A large scale agglomeration of matter, in which may be contained 100 billion stars or more, held together by gravity and usually centered around a massive hub of stars. Galaxies come in a variety of forms, some spiral, some elliptical, some irregular in shape.

Gas giant

A very large planet which is composed largely of gas—mainly hydrogen and helium. Jupiter, Saturn, Uranus, and Neptune are the Solar System's gas giants. They have no solid surface.

Geocentric

Having the Earth at its center. The geocentric theory of the Universe postulated that everything in the cosmos revolved around the Earth.

Globular cluster

A collection of hundreds of thousands of individual stars, all held together in a vast sphere by their mutual gravity.

Glossary

Heliocentric
Having the Sun at its center. The heliocentric theory postulated that the Earth and planets revolved around the Sun.

Light year
The distance traveled by light in one year. At a velocity of 184,000 miles per second, light travels around six trillion miles in a year.

Local group
The cluster of more than 30 galaxies to which our own Milky Way belongs. The nearby Andromeda Galaxy is its largest member.

Magnitude
The perceived brightness of a celestial object is called its apparent magnitude. Its real brightness, called absolute magnitude, takes into account the object's distance.

Meteor
A flash of light caused when a meteoroid burns up on entering the upper atmosphere.

Meteoroid
A small lump of rock in space. If it survives all the way down to the Earth's surface it is called a meteorite.

Milky Way
The name given to our home galaxy. The distant stars in the galactic plane can be seen in a misty band encircling the sky.

Moon
The Moon is the Earth's only natural satellite. Satellites around other planets are also referred to as moons.

Nebula
A cloud of interstellar dust and gas. It may shine by reflecting the light from nearby stars, or by emitting its own light. Dark nebulae appear silhouetted against a brighter background. Old stars are sometimes surrounded by planetary nebulae, well-defined shells of puffed-off gas.

Nuclear fusion
Under high temperatures and pressures, energetic collisions between atomic nuclei form heavier nuclei, and a large amount of energy is released. The fusion process powers the Sun and other stars.

Parallax
The change in an object's apparent position, with respect to more distant objects caused by a change in viewing angle. Nearby stars exhibit a measurable parallax, allowing their distance to be ascertained.

Redshift
The light from rapidly receding objects, such as distant galaxies, is stretched out into longer

wavelengths, toward the red end of the spectrum. Redshift increases proportionately with the distance of galaxies.

Planet

A large non-stellar object in orbit around a star. The Sun has nine major planets.

Satellite

Any object in orbit around a larger body. Most planets have satellites.

Solar System

Our cosmic backyard, containing the Sun and everything that orbits the Sun, including the planets and their satellites, asteroids, and comets.

Star

A huge ball of incandescent gas shining by nuclear fusion. The Sun is a star.

Sunspot

A slightly cooler region on the surface of the Sun which appears dark against the brighter background.

Supernova

The catastrophic explosion of a giant star at the end of its life.

Telescope

An instrument which collects and focuses electromagnetic radiation —from long wavelength radio waves, through visible light to short-wavelength gamma rays.

Optical telescopes produce magnified images of distant objects by using lenses and/or mirrors to collect and focus light.

Universe

Everything we know about—the entire shebang. The Universe is thought to have been created around 13.7 billion years old by the explosion of a primeval atom.

Variable star

A star whose apparent brightness fluctuates over time, either through being eclipsed by an orbiting companion or through changes in its size and/or the level of its light output.

Need to know more?

Joining a local astronomy club or society is a great way to learn more about your Universe. Most large towns and regions across the globe have their own astronomical organizations. A good way of seeking them out is to ask at your local library, scan the local newspapers, or look on the Internet for sources of information.

Societies

American Astronomical Society
2000 Florida Avenue NW, Suite 400
Washington D.C. 20009-1231
http://www.aas.org

The Astronomical League
www.astroleague.org
The Astronomical League website features stacks of useful information and contact details of most astronomy clubs across the United States.

The Astronomer's League National Office
9201 Ward Parkway Suite 100
Kansas City, MO 64114
aloffice@earthlink.net

Association of Lunar and Planetary Observers (ALPO)
www.lpl.arizona.edu/alpo

American Lunar Society (ALS)
otterdad.dynip.com/als

Recommended software

Starry Night
www.starrynight.com
Astronomical simulation programs aimed at all levels of expertise, with excellent tutorials on space and the Universe.

Deep Space Explorer
www.starrynight.com
Explore the wonders of the known Universe in three dimensions. A superb program from the Starry Night folk.

NASA Worldwind
worldwind.arc.nasa.gov
Google Earth
earth.google.com
Explore our amazing planet in breathtaking detail—and for free!

Websites

Astronomy Picture of the Day
antwrp.gsfc.nasa.gov/apod
Each day a different image of the Universe is featured, along with a brief

explanation written by a professional astronomer. Addictive!

Eric Weisstein's World of Astronomy
scienceworld.wolfram.com/astronomy
A great reference site, full of explanations about things astronomical.

The Hubble Heritage Project
heritage.stsci.edu
Features a searchable set of galleries containing Hubble Space Telescope imagery.

Space.com
www.space.com
The very latest space news and views.

Places to visit

Smithsonian National Air and Space Museum
Home of the original Spirit of St Louis and the Apollo 11 lunar module Columbia.
Independence Avenue at 4th Street SW, Washington, DC 20560
http://www.nasm.si.edu
Also located on this site is the Albert Einstein Planetarium
www.si.edu/planetarium/site.html

The Steven F. Udvar-Hazy Center
Houses, amongst many famous spacecraft the Space Shuttle Enterprise

in the S.McDonnell Space Hangar
14390 Air and Space Parkway, Chantilly, Virginia 20151
www.nasm.si.edu/museum/udvarhazy

Kitt Peak National Observatory
The world's largest collection of optical telescopes. Choose from nightly observing programs to gain hands-on astronomy experience or view the Kitt Peak webcam from the comfort of your own home, featuring images updated every minute plus a virtual tour of the observatory.
Tohono O'odham Reservation, Tucson, AZ
www.noao.edu

Hayden Planetarium, Rose Center for Earth and Space
Part of the American Museum of Natural History.
West 81st Street and Central Park West, New York, NY 10024-5192
www.amnh.org/rose

National Solar Observatory
Sacramento Peak, PO Box 62, Sunspot NM 88349-0062
www.nso.edu

Further reading

Books

Dunlop, Storm, *Atlas of the Night Sky* (HarperCollins, 2005)

Dunlop, Storm, *Night Sky* (Collins Wild Guide) (HarperCollins, 2004)

Grego, Peter, *Moon Observer's Guide* (Philips, 2004)

Grego, Peter, *Need to Know? Stargazing* (HarperCollins, 2005)

Levy, David, *Skywatching* (HarperCollins, 2005)

Rees, Martin, *Encyclopedia of the Universe* (HarperCollins, 2001)

Ridpath, Ian, *Collins Gem—Stars* (HarperCollins, 2004)

Ridpath, Ian, *Encyclopedia of the Universe* (HarperCollins, 2001)

Ridpath, Ian, *Norton's Star Atlas* (Pi Press, 2004)

Ridpath, Ian, *The Times Universe* (HarperCollins, 2004)

Ridpath, Ian, & Tirion, Wil, *Collins Pocket Guide to Stars and Planets* (HarperCollins, 2000)

Rukl, Antonin, *Atlas of the Moon* (Sky Publishing, 2004)

Magazines

Astonomy, Astro Media Corp., Milwaukee, WI

Sky and Telescope, Sky Publishing Corp., Cambridge, MA

Index

2003 UB313 87, 115

A
Alpha Centauri 15, 19, 138, 139, 140
Ancestors 24–5, 31
Andromeda galaxy 10–11, 17, 20, 64, 145, 166–7
Ariel 111
Aristarchus 37, 38
Aristotle 37
Asteroid belt 57, 86, 100, 114, 117
Asteroids 52–3, 86, 88, 90, 97, 98, 100–1, 102, 114, 117
Astrology 25, 26–7, 29, 35
Astronomers 174, 175, 176, 178, 180
Astronomy 174–81, 177, 178, 181
    gamma rays 180
    microwave 177
    radio 178–9
    ultraviolet observations 180
    x-rays 180–1

B
Bell-Burnell, Jocelyn 178
Big Bang 16, 66–7
Big whack theory 78–9, 114
Black holes 131, 176, 180
Brahe, Tycho 39, 40

C
Callisto 80, 105
Cassini, Giovanni 45, 54
Cassiopeia 15
Centaurus 10, 19
Cepheids 134, 174
Ceres 86, 98, 100
Chamberlin, Thomas 56
Charon 80, 114
Charon
China 34–5
Comets 25, 26–7, 34, 35, 46, 47, 52, 53, 57, 86, 87, 88, 102, 116–7, 119, 120, 121
Constellations 10, 15, 16, 19, 24, 28–9, 30, 31, 34, 35, 42, 60, 124, 128, 129, 130, 132, 134, 136, 139, 140, 141, 144, 145, 146, 147, 148, 161, 162, 167, 168, 169
Copernicus, Nicolaus 38–9
Cosmology 28–9
Cosmos 8–9, 36, 40, 58, 64, 67

D
Darwin, George 78
Deimos 97

Dwarf satellite galaxies 164–5
    Canis Major 164
    Carina 164
    Draco 164
    Fornax 164
    Leo 164
    Sagittarius 164
    Sculptor 164
    Sextans 164
    Ursa Major 164
    Ursa Minor 164

E
Earth 12–13, 14, 18, 39, 68–77, 80, 81, 86, 87, 88, 89, 90, 92, 93, 94, 100, 101, 102, 110, 112, 114, 117, 120, 134, 140, 142, 174, 176, 178, 181
Eddington, Arthur 10
Egypt 30–1
Eratosthenes 37
Eudoxus 36
Europa 105

G
Galactic clusters 168–9
    Superclusters 169
    Virgo cluster 168–9
Galaxy 10, 14, 15, 16, 17, 20, 21, 58, 59, 62, 63, 64, 65, 126, 129, 130, 138, 142, 144, 145, 146, 147, 154–61, 174, 177, 180
Galileo Galilei 41, 42, 55, 58
Galle, Johann 55
Ganymede 80, 105
Graham, George 47
Greece 36–7, 58

H
Halley, Edmond 46
Heavens 8–9, 24, 25, 26, 28, 30, 34, 38, 42, 58, 62
Herschel, William 55, 58–9, 149
Hewelke (Hevelius), Johannes 43
Hewish, Antony 178
Hipparchus 37
Hooke, Robert 54
Horsehead Nebula 146–7
Hubble Space Telescope 120–1, 150–1, 170–1, 180, 181
Hubble, Edwin 64–5
Huygens, Christiaan 44–5, 51, 55, 107
Hyades 143

Index

I
Infinity 8, 154–61
Io 80, 104–5

J
Jupiter 14, 80, 45, 54–5, 86, 102–5, 117

K
Kant, Immanuel 56
Kepler, Johannes 40, 41
Kuiper Belt 114–5

L
Laplace, Simone 56
Large Magellanic Cloud 162–3, 164
Leavitt, Henrietta 63
Leverrier, Urbain 55
Lo-hsia-Hung 34
Lowell, Percival 50

M
Mars 14, 45, 50, 86, 94–9, 117
Megaliths 32
Mercury 14, 48, 86, 90–1
Messier, Charles 53, 59
Milky Way 15, 16, 17, 20, 30, 41, 58, 62, 63, 64,
        126, 134, 138, 142, 145, 146, 147, 148,
        159, 160, 162, 163, 164, 165, 166, 167,
        168, 170
Mira 136–7
Miranda 111
Moon 12–13, 18, 25, 26, 28, 29, 42–3, 49, 78–83,
        86, 87, 90, 104, 114, 141, 174, 175, 181
Moulton, Forest 56

N
Neptune 14, 55, 86, 87, 112–3, 114
Newton, Isaac 46

O
Oberon 110–1
Oort Cloud 14, 15, 19, 116, 117, 119
Orion 30, 31, 124, 128, 132, 146, 149
Orion Nebula 146, 149
Orreries 47

P
Parsons, William 62
Penzias, Arno 67
Philolaus 38
Phobos 97
Piazzi, Giuseppe 52
Planet X 87, 115
Pleiades 143, 151
Pluto 14, 55, 80, 87, 114–5

Ptolemy 37, 38
Pythagoras 36, 38

Q
Quasars 154, 157

R
Riccioli, Giovanni 43
Roemer, Ole 45

S
Saturn 14, 45, 55, 80, 86, 106–9
        moons 108–9
Schiaparelli, Giovanni 50, 51
Shapley, Harlow 63
Shih-Shen 34
Sirius 141
Small Magellanic Cloud 63, 162, 163, 164
Solar system 14, 15, 19, 40–1, 45, 47, 48, 53, 55,
        56, 57, 70, 73, 80, 86, 87, 88, 90, 94, 96,
        102, 105, 106, 112, 115, 116, 117, 119, 120,
        138
Spectroscopy 126
Stars 10–11, 24, 26, 29, 30, 31, 32, 34, 35, 36, 37,
        38, 39, 41, 43, 46, 52, 56, 58, 59, 60, 61,
        62, 63, 64, 65
        Globular clusters 144–5
Sun 10–11, 13, 25, 26, 28, 38, 58, 59, 86, 88–9, 93,
        98, 100, 107, 110, 112, 114, 115, 116, 117,
        119, 124, 126, 127, 128, 130, 131, 132, 138,
        139, 140, 141, 142, 147, 154, 155, 159, 178,
        181
        Solar activity 88
        Solar cycles 89
Supernovae 130–1, 137, 175, 180

T
Telescopes 174–5
Titan 80, 106, 108–9
Titania 110
Tombaugh, Clyde 55
Triangulum Galaxy 20, 64, 167
Triton 112

U
Umbriel 111
Uranus 14, 55, 86, 87, 110–1

V
Venus 14, 26, 29, 49, 86, 92–3

W
Whirlpool Galaxy 62
Wilson, Robert 67

Acknowledgments

I dedicate this book to my wife Tina and my daughter Jacy, for providing a Universe of love and support—my gratitude is boundless, and I wish to reflect and focus that love to you both.

The staff at Collins have been tremendously helpful, patient, and enthusiastic about this book from the very start—my thanks to Helen Brocklehurst, and special thanks to Julia Koppitz for seeing the book through from start to finish. Finally, thanks to Focus Publishing and David Etherington for his professional and positive attitude. You've all done a great job!

Peter Grego